J. E. C. PETERS

Discovering Traditional Farm Buildings

D0767712

SHIRE PUBLICATIONS LTD

Contents

Cover: *Slad Barn, Hawling, Gloucestershire. (Photograph by J. E. C. Peters)*

Published in 2003 by Shire Publications Ltd, Cromwell House, Church Street, Princes Risborough, Buckinghamshire HP27 9AJ, UK. (Website: www.shirebooks.co.uk)
Copyright © 1981 by J. E. C. Peters. First published 1981; reprinted 1986; reprinted and updated 1991 and 2003. Number 262 in the Discovering series. ISBN 0 85263 556 7.

Printed in Great Britain by CIT Printing Services Ltd, Press Buildings, Merlins Bridge, Haverfordwest, Pembrokeshire SA61 1XF.

Introduction

Farm buildings form an important element in the country landscape. Whilst their detailed design changed only gradually in the past, the rate and type of change is now very much greater as a result of the considerable alterations in farming methods since the Second World War, with greatly increased mechanisation and automation, and the combination of farms. In consequence the traditional buildings have become largely redundant; a great many have been or are being converted to dwellings, some to other uses. If they are kept with the farm they may be modified, sometimes drastically, demolished or left to decay.

Much has been written on the development of the farmhouse and its regional variations, but similar work on farm buildings is a more recent development. Some books and articles have been published on localities, individual buildings or specific aspects of design. Much, however, remains to be recorded before the evidence completely disappears.

Farm buildings provide valuable material evidence for agricultural history. They reflect the considerable regional differences in types of farming and, overlying these differences, in methods of housing and threshing the crops or housing and feeding the livestock. As farming methods changed, so the farm buildings may have been changed to adapt to the new ideas. The speed of change varied, however, so that a building design considered out of date in one area might still be fashionable in another.

The design of the buildings was also affected by the local building tradition. During the nineteenth century this was to be overlaid by a national style, the speed with which this happened varying from one area to another. In addition some estates had their own distinctive style of building.

To study the local building tradition and its dissolution is beyond the scope of this book; it will concentrate on examining the characteristic features and plans of the different types of farm building before 1880, so that each may be recognised, as a barn, stable, cartshed or cowhouse. Such variations as are currently known will be noted. The intention is to look at each building in the way that it is approached, first its external details, then the internal and finally the plan. The farmhouse is a subject on its own and will not be included.

The book is largely based on the author's field research in various parts of Britain and on his study of farming textbooks of the eighteenth and nineteenth centuries, as well as later, more historical works. He would like to thank the various members of the Vernacular Architecture Group and others who have helped with comments and discussion on the subject, in particular Dr N. W. Alcock, Dr R. W. Brunskill, Mr R. Harris, Mr L. F. J. Walrond, Mr W. J. Smith, Mr M. J. Hill, Mrs Beaton, Mr I. Homes, who helped with the section on hopkilns, Mr J. Carter, who introduced him to Suffolk farm buildings, and Mr J. McCann for his expertise on dovecotes. His thanks are also due to the partners of ASTAM Design Partnership, Gloucester, and Mr B. J. Ashwell for their encouragement, and to all those farmers who have permitted him to examine their buildings, without whose permission this work would not have been possible. Finally, he wishes to acknowledge the continuing encouragement and support of his wife.

a.

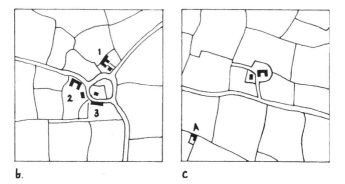

b.   c

Fig. 1. Settlements: a, village; b, three-farm group; c, isolated farm with field barn, A.

4

1. The farmstead

The farmstead may be defined as the farm buildings and house looked at as a group. It may help our understanding of the individual buildings which form it to look first at where it was built and how it was arranged.

Farmsteads are found in nucleated or closely grouped settlements (either villages or hamlets), in small groups of two or three farms, or they are isolated. Their location is the result of centuries of development and change. Those in villages, and to a lesser extent in hamlets, are the survivors of a larger or possibly much larger number, depending on how far back you look. A settlement of two or three farmsteads, or even an isolated one, could be the shrunken remains of a village or hamlet: in a few cases the first may represent an isolated farm which has been later subdivided. The isolated farmstead normally represents one of two developments. The first is the occupation of previously unfarmed land by a single farm, at some time from before the Norman Conquest to the late nineteenth century. The second is the movement of a farmstead out of a village or hamlet, following enclosure of the open fields or the redistribution of land enclosed at an earlier date. Some background research will invariably be necessary to understand the development of the site and of the farm.

Sometimes farm buildings were built away from the central farmstead. Generally these did not form a fresh farmstead, lacking some necessary buildings, usually the stables. These non-farmstead groups are of two types. The first, the field barn, is the more widespread. It consisted of a threshing barn, usually with a shelter shed and foldyard for loose cattle. They were built to serve land lying some distance away from the farmstead, or which could only be reached from it by steep hills. The intention was to save carting the corn to the farmstead for threshing and carting manure back again, the cattle converting the straw to manure at the field barn. They seem to have been connected with mixed farming, cattle being kept to fatten. They first appear in the seventeenth century, ceasing to be built about the middle of the nineteenth, when the use of portable threshing machines meant that ricks could be built and threshed in the fields and a barn was no longer needed. Many have now disappeared.

The second type of detached building was the field house, found in some hilly or mountainous pastoral areas. It is a building erected to house the hay grown in the nearby fields, with a small cowhouse in one or both ends. Like the field barn, it was built to save carting produce back to the farmstead.

The number and size of the buildings of the farmstead were

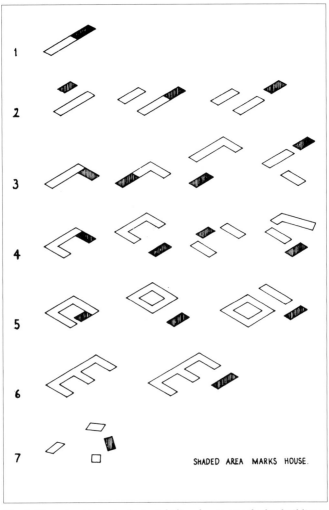

Fig. 2. Farmstead layouts: 1, straight line, house attached to buildings; 2, parallel; 3, L plan; 4, U plan, the last irregular from awkward site or buildings of more than one period; 5, enclosed square; 6, multiple yards; 7, irregular (after R. W. Brunskill).

affected not only by the size of the farm but also by the type of farming practised. On an arable farm with only a few cattle kept to make manure, there would be little in the way of buildings for cattle, but there would be one or more barns, stables and shelter for carts. On a mixed farm there would be a number of buildings for cattle in addition, while on a pastoral farm the stables would be small and there would be little accommodation for crops. Sheep had very little effect on the farmstead, however big the flock, as (except in Herefordshire) buildings were only very rarely provided for them.

The actual pattern that the buildings form is to some extent dictated by the number of buildings available: it is also affected, at least in later examples, by a desire to shelter the yard. The drawings in fig. 2 show a simple analysis of the common types. Types 1 and 2 are found on small farms with but few buildings. More were generally needed for type 3, which is the first to try to shelter a yard. The multiple yards of type 6 are found generally only in large nineteenth-century farmsteads; the covered yard is really a variation of 4 or 5. There are other, complex nineteenth-century farmsteads which are not readily covered by such diagrams. Regularity of plan may be achieved, even though the farmsteads were built over a long period; the last in 4 is a slightly irregular result. The final type is the haphazard plan, which usually resulted from buildings going up in the next available space, without proper consideration of their relationship to each other. It is probably the oldest plan type.

The relationship of the individual buildings within the block layout was important if labour were not to be wasted. Threshed straw had to be taken from the barns to the cattle or horses for use as feed or litter; later other feed was distributed from the barn. Cattle being the principal users of straw, it was important to have them near the barn; the stable could then either be beyond the cowhouse or adjoin the barn on the other side. The third alternative was to put the stable next to the barn, with the cowhouse beyond; although much less satisfactory than the other two, it is occasionally found.

Sometimes, for economy in building costs and perhaps to avoid spreading the buildings out too much, some were combined into a single building. One of the types found most frequently is the barn with lean-tos built against it, forming cowhouses, stables, cartsheds or, perhaps most often, sheds for yard cattle (fig. 3a). The cost of one wall was saved and, by putting the buildings close to the barn, they could more easily be supplied with straw. The other common type was where the cowhouse, stable or cartshed was put under part of the barn; the affected area was, in consequence, raised above the threshing floor (fig. 16). This first appeared in the

7

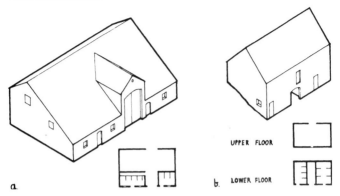

Fig. 3. Combination buildings: a, lean-to to barn; b, bank barn.

seventeenth century, in some cases to take advantage of the slope of the ground.

The bank barn was a different response to a sloping site. The barn was built along the slope, being entered at ground level on the uphill side (fig. 3b). On the opposite, downhill side it was at first-floor level, with the cowhouse, stable and sometimes cartshed tucked underneath. This economical building is principally found in the Lake District; it first appeared in the second half of the seventeenth century, the last known example being dated 1905. It is also found in a few other parts of Great Britain, but is then apparently a nineteenth- or possibly late-eighteenth-century phenomenon. It is not unlike the Pennsylvanian bank barn.

The last type of combination building is the bastle house, restricted almost entirely to the southern side of the England-Scotland border. It consists of a dwelling house on the first floor, with a space underneath for cattle, horses and sheep. It was a defensive building, erected between the mid sixteenth and mid seventeenth centuries. Some later ones are found in County Durham.

The final aspect of the farmstead to be examined is the relationship of the farmhouse to the farm buildings. The house may be attached to some or all of them. Harrison, writing in 1577, considered this a northern characteristic, but it did then also apply to the West Country. The type of building to which the house is attached is significant. The dwelling and cowhouse adjoin in the *longhouse* type, forming a straight line as in fig. 4a. The origins of this arrangement have been the subject of much discussion. In early examples people and cattle used the same doorway, which opened into a passage which separated their accommodation. In later examples the two buildings merely shared a gable, with no intercommunication. Where the barn adjoined the dwelling it is called a *laithe house*; it often had the stable and cowhouse built beyond the barn as part of the same range (fig. 4b). This type is found in the Pennine area of Yorkshire, spreading into the Lake District. In the Yorkshire area it dates from the seventeenth century until towards the end of the nineteenth, most examples being of the nineteenth century. The house could also be attached to other types of building, or at right angles to those already noted, but these do not form distinctive types.

The attachment was often part of the original design: sometimes, however, a detached house became connected to the farm buildings by the building up of the gap. In most cases attached buildings occur on small farms. Some exceptions appeared in the nineteenth century on large farms, however, the intention being to create an architectural effect by linking house and buildings in one range.

Whether the farmyard could be seen from the house was also important. In some cases the front or back overlooked the yard, usually with at least a very narrow garden or yard between. However, so close a contact with animals and manure was not always desired, especially by the farmer's wife. One extreme solution was to move the farm buildings well away from the house, even out of sight of it. The alternative was to make them ornamental: such examples are usually sited a little way from the house. Both methods are to be seen at Acton Scott, Shropshire. The buildings were ornamented when erected in 1769, but early in the nineteenth century trees were planted to hide them from the house. Where the farmstead is away from the house a labourer's cottage may be part of the group, so that there would always be someone about the buildings for security.

Fig. 4. House attached to buildings: a, longhouse; b, laithe house.

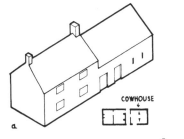

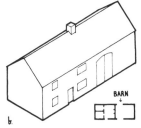

2. The barn

In most traditional farmsteads the barn was the central point to which the other buildings related and was often the largest of them. For this reason it is the one to be examined first. There has been a fairly recent trend of using the term 'barn' to describe any farm building; this has no historical basis, and is totally incorrect. Each building has its own particular use and characteristics, as will be shown below; to call all of them a 'barn' can only cause confusion.

The barn was a building for housing and threshing the corn and pulse crops grown on the farm: on the one hand wheat, oats, barley, rye, or a mixture of two or more, on the other hand peas and beans. It might also house hay and threshed straw. This definition covers all barns built before the early nineteenth century and many, if not most, thereafter. Generally they had at least one threshing floor, with bays for housing the crops opening off it (fig. 5). The threshing floor always ran across the barn, never along the length, as in parts of mainland Europe. The exceptions to the definition had no provision for threshing but were used for preparing the feed for the livestock on the farm, with space for housing hay and straw.

The size of barns varied greatly, but in spite of this there were nearly always only one or two threshing floors; very few barns seem to have had more and these are all medieval in date. (Not all medieval barns were necessarily tithe barns and most which were also served as the barn for a farm.) The most obvious factor may seem to be the size of the farm, but its effect was modified by the type of farming. A farm of 100 acres which was largely arable would produce much more in the way of crops than one of the same size which was largely pastoral or producing milk and dairy products. The type of farming could, like some of the other factors, change from period to period. The most important factor was whether all the crops were housed in the barn, or whether some or all were kept in ricks, being moved to the barn for threshing. The practice in the Middle Ages was to house as much as possible of the grain crops, preferably all of them, in the barn, using ricks only as a last resort when the barns were full. This naturally made for very large buildings. The method of harvesting could also affect the size, in that corn bound into sheaves needed less space than corn harvested loose. In contrast, in Scotland and Northumberland, by the early eighteenth century, the bulk of the crops was housed in ricks; the barn in consequence was small, having only space for housing one rick at a time and a threshing floor.

The practice of storing all the crops in barns, and so of building large ones, continued well into the nineteenth century in the south of England and in East Anglia, in spite of most contemporary

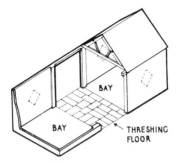

Fig. 5. Cut-away view of barn.

writers advocating the use of ricks. The reasons given for building large barns were that the grain was better housed in barns and was always at hand to be threshed; it was considered to be cheaper because the cost of thatching ricks was avoided, and double handling in moving the ricks to the barn for threshing was saved. These arguments ignored the capital cost of the barns and the costs of maintenance. Availability of finance is a fourth factor. Whilst the small farmer might have difficulty in building even a small barn, at the other end of the scale there may have been some overbuilding for show, although this was probably rare.

An alternative to building one large barn was to build two or more smaller barns to give the same capacity: the capital cost could then be spread over a longer period. As the use of ricks became more usual, the number of barns might be reduced, or one new one built and the old, perhaps by then in a poor state of repair, demolished or, on occasion, converted to some other use.

The barn was filled with harvested crops, leaving only the threshing floor clear. The crops were stacked up to the ridge, or close to it, so using nearly all the roof space. The use of half the barn for threshed straw, filling only one side of the threshing floor with unthreshed corn, seems to be a late development, connected with the use of ricks rather than barns for housing crops, or with faster threshing by machinery rather than by flail. Whilst the names of some barns, such as the wheat and barley barns at Cressing Temple in Essex, imply the use of a barn for one crop only, documentary records from the Middle Ages to at least the early eighteenth century show that in practice barns were used for two, three or more crops at once. With two it might simply be one on each side of the threshing floor, but on occasion two or three different crops might be housed on one side.

The characteristic features of a traditional threshing barn will now be examined, beginning with the external ones.

11

Doors and porches

The doors to the threshing floor are generally the most prominent external features. Where they have been built up or reduced in size the original outline can usually still be made out.

The most common type is the large, wide doorway, closed by double doors: these might be subdivided horizontally to give three or four sections. The doorway was high to admit laden waggons or carts to unload from the threshing floor into the bays; its height also provided light for threshing and air for winnowing the husks from the threshed grain. In some barns the floor could only be approached from one side, there being a blank wall opposite (fig. 6f). This prevented a through draught for winnowing and meant that carts had to back out.

An alternative to a high, wide doorway was one of the same width, but only half the height (fig. 6b). It had the advantage of being cheaper to make and maintain, as various nineteenth-century writers noted; set opposite high doors, it could be used to let out the cart after unloading. Sometimes such doors were used at both ends of the threshing floor; they were then too small to provide a natural draught for winnowing, but this did not matter if one were induced by a fan or machinery were used. The winnowing machine seems to have appeared during the second half of the eighteenth century,

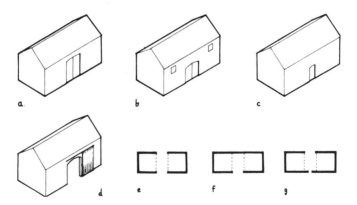

Fig. 6. Door sizes for a barn: a, high and wide; b, low and wide, with pitch-holes; c, small; d, sliding; e, plan with wide doors, high or low; f, wide door on only one side; g, small door opposite wide.

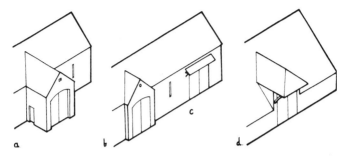

Fig. 7. Porches: a, large, with small door in side; b, slight projection; c, canopy; d, in an aisled building, based on Berkshire examples.

and the sole use of such doors may be connected with this. The main disadvantage was that the barn had to be filled by barrow or through pitch-holes; by the latter method it took twice as long to unload a cart, and, being outside, it was not protected from the weather.

A small door was sometimes used opposite a large one, as in bank barns in the Lake District (fig. 3b); this combination was said to improve the draught for winnowing. Used at both ends of the threshing floor, it had the same effect as the wide, low door. This combination was beginning to be used in Yorkshire in the late eighteenth century, where barns were small and ricks were used to house the crops. Small doors might additionally give access direct to the storage bays, being useful for cleaning them out when the barn was empty or if the bays were used for housing sheep at shearing time (fig. 11).

Sliding doors appeared in the 1840s but were never very common, being more expensive to install than hinged ones and more likely to stick. Their advantage was that they did not catch the wind and did not exert the same strain on their supports. They were used for large, and wide, low doorways alike (fig. 6d, plate 8).

Porches are found in many surviving medieval barns and in some later ones, particularly in parts of the south of England (fig. 7). They were useful in increasing the size of the threshing floor and in protecting it in inclement weather when the doors had to be open. Laden carts or waggons could shelter in them if the threshing floor were already occupied by others unloading. Where there was not a separate cartshed, the porch could be used as one. Some

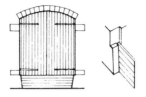

Fig. 8. Barn doors, showing the lift.

porches had a small side door permitting access without opening the large ones. In a few cases there was a room over the porch, the earlier examples probably used by the farm bailiff, as at Bredon Tithe Barn in Worcestershire. Examples from the eighteenth and nineteenth centuries in the Cotswolds were used as granaries.

In aisled barns with low side walls or where lean-to buildings give a similar external effect, the porch might not project forward from the low wall, becoming an outsize dormer to give height for laden carts (fig. 7d). Others in high-walled barns with little or no projection were purely ornamental.

The barn doors might stop between 1 and 2 feet (305-610 mm) above the ground, with a separate series of boards below (fig. 8). These were held in place by a groove in the bottom of the door jamb, which projected slightly for the purpose. In timber buildings the groove would be formed by fixing battens to the door frame. The lift, as it is called, enabled the doors to clear the manure in the yard and kept pigs off the threshing floor when the doors were open. They would also catch grain bouncing with the force of flail threshing. A cloth might be hung from a higher level to supplement the lift or to perform this function in its absence.

Pitch-holes

Window-like openings, covered by wooden shutters, were sometimes provided in the sides and ends of barns (fig. 6b). They were pitch-holes, used for pitching corn or hay into the barn off a cart standing outside. They could also be used to give light and air if the barn were not full; those high up in the gables may have been intended only for this purpose. The importance of these openings where the doors were too low for laden carts has already been noted, but they were not restricted to such buildings. In shape they were generally just off square. After about 1825 circular ones began to appear: it was considered that with the changed shape the corn was less likely to catch on the edges of the opening, shedding some of the grain.

Air vents and owl holes

Air vents were provided in the walls of barns to prevent the crops inside from becoming mouldy as a result of any damp left in them when they were housed (fig. 9). Vents also provided some light when the barn was empty, but this was a secondary benefit. The amount of ventilation provided varied considerably, partly affected by the materials of which the barn was built; it was easier to provide overall ventilation in brick and timber framing than in stone.

Air vents in stone barns were normally slits or single holes (fig. 9a, d), often widely spaced, sometimes more obviously arranged in rows, particularly in later examples. The holes were at first square, but a triangular form appeared during the mid eighteenth century.

What provision was generally made in medieval timber-framed barns is not clear: some woven panels have survived in a few Gloucestershire, Herefordshire and Worcestershire examples, but it is not known if their use was more widespread (fig. 9c). No

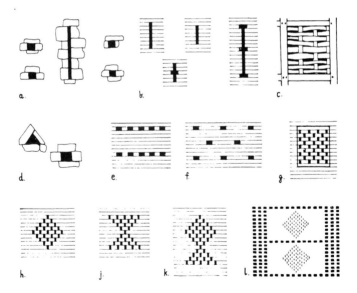

Fig. 9. Air vents: a, combination of slit and single holes, stone wall; b, various slits in brickwork; c, woven panel; d, single holes in stone; e, f, rows in brick wall; g, grouped into a panel; h, j, k, geometric shapes; l, pattern.

15

special provision was needed with weatherboarding, as the timbers did not fit closely together. With brick nogging various patterns could be formed by omitting four to six headers. The high brick plinth in some eighteenth- and nineteenth-century barns might also incorporate air vents.

The greatest variation of type was found in brick barns owing to the ease with which patterns could be formed by omitting a half-brick or less without affecting the structural stability of the walls. Slits seem to have been the earliest type, borrowed from stone barns: they were generally found in rows. In some areas, as in Nottinghamshire, they continued to be built well into the nineteenth century. Half-brick holes could readily be arranged in rows; they had appeared by the late seventeenth century and could provide good overall ventilation. Geometric shapes appeared a little later but are perhaps the most widespread type. They varied widely in size and were found singly or in combination with each other or with variations on fig. 9e, f: fig. 9l is an example of about 1860 from Seighford in Staffordshire. A final alternative was to group the vents into a series of windows (fig. 9g). In some areas, however, such as Glamorganshire and south-west Yorkshire, very little ventilation was provided.

Not to be confused with air vents is the *owl hole*, a nearly square or circular opening of about 6 to 9 inches (152-228 mm) across, set high up in the gable of the barn. These first appeared at the end of the seventeenth century to let owls into the building to catch mice (fig. 7a).

Internal subdivision

The most important point internally was the method of separating the threshing floor from the bays (fig. 10). Before the seventeenth century, and in most of England and Wales thereafter, there was no subdivision, the barn being open from end to end. Even in areas where subdivision appeared it was not found in all barns. The only exception before the seventeenth century was in some timber-framed aisled barns which had a tie between the outer wall and the foot of the aisle post.

In the seventeenth century, in timber-framed barns in the west Midlands and along the Welsh border, the practice began of dividing up the barn bay by bay, the normal pattern being as in fig. 10b. The bracing reinforced the walls against wind and the pressure exerted by the crops housed inside and might also reinforce the doorposts against the weight of the doors. It helped to keep the loose crop or straw inside the bay and separate them from the threshing floor. It was later realised that the cill wall was useful in catching grain bouncing with the force of flail threshing and in forming pens to

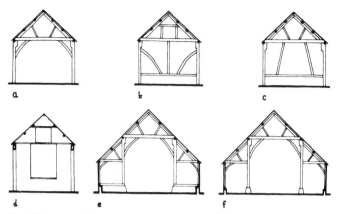

Fig.10. Internal divisions in a barn: a, open; b, timber-framed; c, late variation of b; d, brick spur walls; e, aisled, tie between foot of main posts and outer wall; f, aisled, without tie.

house sheep before shearing or for sorting livestock. Boards were sometimes added either side of the threshing floor to increase the height for the last two purposes, and a small door might be cut under or, with less structural logic, through the tie, to let in the livestock. This framing lasted, with variations in the form of the bracing, until the second half of the nineteenth century in Worcestershire and Herefordshire. Such timber framing is sometimes found in brick or stone barns in the same area, presumably because timber framing was still the normal form of construction or was only just being replaced.

A mid-eighteenth-century development in brick and stone barns was to replace the timber bracing with spur walls, effectively internal buttresses; they were normally linked by a low wall (fig. 10d). This variation is found in west Staffordshire, north Worcestershire, north-east Cheshire and south Lancashire, including the Fylde. In Cheshire and Lancashire there were apparently no timber-framed precursors. The use of low walls on their own, without spur walls or timber bracing, had a much wider distribution.

Cornholes and cupboards

A small room may be found opening off the threshing floor on one side, very occasionally balanced by one on the other side. This is the cornhole, a mid-eighteenth-century development so far known only in Staffordshire and Suffolk, with a few in east Sussex. In

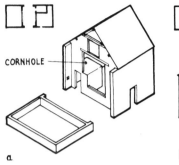

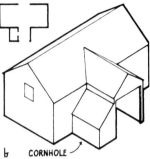

CORNHOLE

a.   b.   CORNHOLE

Fig. 11. The cornhole.

Staffordshire it is always inside the barn, as in fig. 11a, between 3 and 4 feet (0.91 and 1.22 metres) wide, 6 feet (1.83 metres) high, usually with plastered walls and flat timber top, although a few have a brick vault. In Suffolk it stood in the angle between the barn and the porch (fig. 11b). It was built to house the mixed grain and chaff after flail threshing until enough had been collected to make winnowing worthwhile. Its provision may have been connected with the use of machinery for this task.

A much less obvious feature is the cupboard, a small recess by the door or in the porch wall found in some brick or stone barns. It held the horn containing grease for the joint in the flail, which had to be used two or three times a day. It may also have housed a lamp. There may also be a series of vertical scratch marks by the door, made by farm labourers engaged in the monotonous work of flail threshing either to tell the time or to count the quantity of work done.

The plan

The plan of the barn is governed by the position and number of the threshing floors, producing five types in traditional barns: these are not affected by the overall length of the barn or by the materials of which it is built. The first and oldest plan is where the threshing floor is in the middle. A single floor may, however, be off-centre or even at the end of the barn. The fourth type comprises barns with two or more threshing floors, and the last type barns where the storage bays are set at a higher level than the threshing floor, with some other, non-barn use beneath. Sometimes two or more barns were built end to end, the barns being separated by a wall with no openings or only a pitch-hole. These are separate buildings, each with its own plan form.

There was considerable variation in the size of barns, as the aerial views in fig. 12 show. This was not necessarily related to the number of threshing floors: Middle Littleton, Worcestershire (c.1290), with two, is smaller than Great Coxwell, Oxfordshire (late thirteenth to early fourteenth century), with one. The largest barn of all, at Beaulieu, Hampshire (thirteenth century), had only one floor. Audmore Farm, Gnosall, Staffordshire, has been included as a nineteenth-century example designed for threshing ricks.

Whilst the variation in size can be defined by the actual length, it can also be expressed in terms of the number of bays, a bay being

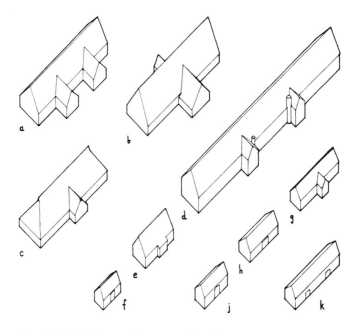

Fig. 12. Aerial views of barns, to show the variation in size (all to same scale): a, Middle Littleton, Worcestershire, c.1290; b, Great Coxwell, Oxfordshire, late thirteenth to early fourteenth century; c, the Wheat Barn, Cressing Temple, Essex, c.1250; d, Abbotsbury, Dorset, c.1400; e, Court Barn, West Pennard, Somerset, fifteenth century; f, Valley Farm, Ubbeston, Suffolk, fifteenth century; g, Great Budbridge, Isle of Wight, eighteenth century; h, Village Farm, Seighford, Staffordshire, 1758; j, Audmore, Gnosall, Staffordshire, nineteenth century; k, Wolgarston, Penkridge, Staffordshire, 1801.

the section of building between two roof trusses. This method was used in many documents from the Middle Ages onwards. It does, however, present slight problems of interpretation, as the length of a bay is not a fixed measurement but could vary, normally for a barn between 14 and 18 feet (4.27 and 5.5 metres). Period of construction was one factor: the longest bays in part of Staffordshire date from the first half of the eighteenth century, up to 20 feet (6.1 metres); thereafter the length dropped to about 10 feet (3.0 metres). Furthermore, not all the bays in one building were necessarily the same length.

Aisled barns

Aisled barns fit into at least three of the plan groups, without affecting their basic definition, so they are considered first. The use of aisles permitted the erection of a much wider building than would otherwise have been possible; aisled barns are normally between 29 and 50 feet across (8.8 and 15.2 metres). A base-cruck truss could span a building of between 30 and 36 feet (9.15 and 11 metres) across internally, but otherwise the maximum width was usually 30 feet, most barns being at least 5 feet (1.5 metres) narrower. Although using aisles might appear a useful way of building a large barn without it becoming excessively long, many such barns are of some length, and some early ones of great size. One recorded in 1155 at Waltham in Essex was 168 feet long by 53 feet wide (51.2 by 16.15 metres). A fourteenth-century one at Frindsbury in Kent was 212 by 34 feet (64.6 by 10.3 metres).

Aisled barns are found in two main areas, both dating back to the Middle Ages. The first, with mainly timber-framed examples, is from Wiltshire and Hampshire across to Suffolk and Kent. The second area is in south Yorkshire and the Pennines and seems to have spread over into mid Lancashire in the early seventeenth century. Scattered examples, whether with timber-framed, stone or brick walls, are found elsewhere. There are some areas, however, where no aisled barns were apparently ever built, such as Staffordshire and the whole of Wales. Barns with only one aisle seem to be a development of the seventeenth century. In some barns the aisles were not used for storage but were shut off for some other purpose such as cowhousing, as in some Yorkshire examples in the seventeenth or eighteenth centuries. These are not strictly aisled barns but combination buildings, and so they do not come within this category.

Flail-threshing barns: plan types

The earliest plan (type 1) to appear was the barn with a single, central threshing floor; it is by far the commonest type and is found

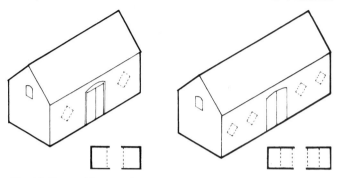

Fig. 13. Barn with central threshing floor, type 1.

throughout Britain (fig. 13). Some of the early examples were very large, having as much as 50 to 100 feet (15 to 30 metres) between the threshing floor and the end wall. Filling such barns presented problems, at least at a later stage. It was also difficult to get at corn stored at the ends, which might well be a different crop from that adjoining the threshing floor. Not surprisingly, many such barns have had additional openings made and sometimes further threshing floors added.

Most extant barns of this plan are, however, much smaller, with only one or perhaps two bays on each side of the threshing floor. This is partly due to the later use of ricks to house the corn, the barn being reduced to a building for threshing, with space for housing only one rick and the resulting straw: for this the small barn with a central threshing floor was admirably suited. With the decline in timber, and so in bay sizes, the nineteenth-century barn with two bays on each side of the threshing floor may be little longer than an earlier barn with only one each side.

The second oldest plan type is the barn with two threshing floors, which had appeared by the late thirteenth century (fig. 14). The increase in the number of threshing floors permitted faster threshing and gave easier access to the interior. It was not just a barn for large farms: a number were built on relatively small farms in central, western Staffordshire in the mid eighteenth century and in Breconshire in the 1770s and 1780s. Some barns which appear from the outside to be of this plan are in fact two barns of the first type built end to end, with an imperforate wall or one with no more than a pitch-hole between. Ashleworth Tithe Barn, Gloucestershire (c.1500), was an example but the dividing wall has been removed. A variation had only two storage sections, one of the threshing floors being thus at the end of the barn.

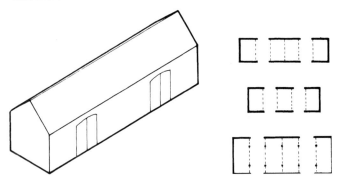

Fig. 14. Barn with two threshing floors, type 2.

A few barns had three or more threshing floors. Some were built to this plan, like the fifteenth-century tithe barn at Harmondsworth, Middlesex; others are the result of cutting additional openings in an earlier long barn. At least two barns are recorded as having had four threshing floors: one was the great barn at Cholsey, Oxfordshire, demolished in 1815; the other was at Canbury House, Kingston upon Thames, where twelve carts could unload at once.

The third plan type had a single threshing floor placed off centre as in fig. 15a. Whilst the reason for this plan is not clear, it is likely that in early examples the whole barn was intended for unthreshed crops, from the desire to house as much as possible. To intend the short end for threshed straw, as some modern writers state, must be a nineteenth-century development, with fast machine threshing and the use of ricks. This plan had appeared by the late fourteenth century and was to continue to be built until the end of the nineteenth. It is widely distributed throughout England and Wales.

Fig. 15. Barn with threshing floor off centre: a, type 3; b, type 4.

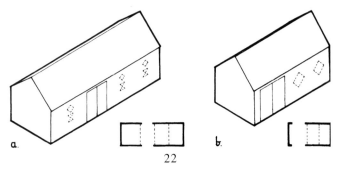

Normally there was a different number of bays on each side of the threshing floor, as in the tithe barn at Church Enstone, Oxfordshire (1382), which has two on one side and three on the other. Sometimes, however, the barn had only one bay on each side of the threshing floor, but the lengths of each were markedly different.

In the fourth type the threshing floor was placed at the extreme end of the barn (fig. 15b). It is not clear when this type first appeared, but examples survive from the early seventeenth century onwards: it was not a very common plan. Some examples were extensions to earlier, larger barns, and so merely show a need for more barn room. Most such barns, however, were, or are now, the only barn on the farm; these were generally small farms or had a small arable acreage. Some nineteenth-century examples on larger farms may have been connected with the use of ricks to store the corn, being built to shelter the threshing machine and house straw, or perhaps only the latter.

The fifth plan type (fig. 16) has part of the storage bays set at a higher level than the threshing floor, usually up a full storey. In a few cases all of the bays are so raised. The space beneath is used for some non-barn function, usually cowhousing, but stabling and cartsheds are also found. The break in level was nearly always at the edge of the threshing floor; only occasionally was it set back a bay. Where cattle were kept underneath, the threshing floor either acted as the feeding passage (plate 8) or had access to it. In some cases advantage was taken of a sloping site to put the cowhouse at the low end of the building, so reducing the difference in levels between the two parts of the barn. The plan was particularly suited for small, probably pastoral farms with a little arable and seems to have appeared by the early seventeenth century. By combining some of the buildings it economised in building costs.

Fig. 16. Barn partly in loft over other use, type 5.

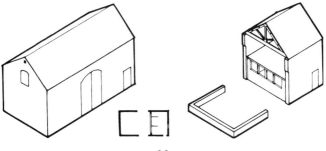

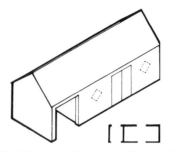

Fig. 17. Covered cartway adjoining barn.

The distribution of this plan is somewhat uneven, with known areas of concentration in north Cheshire, Lancashire south of the Lune, the Yorkshire Dales, Radnorshire and part of north-west Wales, where a survey found over half the barns to be of this type. The Lancashire examples differ from those elsewhere in that in many cases the cowhouse is wider than the barn, producing a T-shaped building (fig. 34f).

Very occasionally a covered cartway was provided adjoining the barn, with doors at one or both ends as in fig. 17. At first sight it appears not unlike another threshing floor, but internally there are solid walls on each side of the passage. It was something of a luxury and might be provided partly for architectural effect, linking two buildings of similar height, as at Acton Scott in Shropshire. It could also serve to shelter a laden cart when the threshing floor was already occupied, or as a grand waggon shed.

Regional types

A number of regional variations in the design of barns are known; in each case the barn will fit into one of the five types already noted, but there are additional features warranting separate notice.

The *bank barn* is the best known and probably numerically the largest variation. It has already been described as a form of combination building particularly suited to sloping sites (fig. 3b). Principally found in the Lake District, scattered examples have been found in some other hilly parts of Great Britain, particularly Devon and north-west Scotland.

Fig. 18a shows another, possibly related, variation, found in east Devon and the adjoining part of west Somerset, apparently dating from the nineteenth century. Here the barn, usually small, may be put entirely at first-floor level, no advantage being taken of the sloping ground. The doors to the threshing floor are both large. The

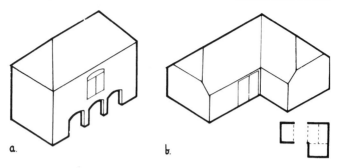

Fig. 18. Variant types of barn: a, Somerset first-floor barn; b, L plan in Wiltshire (Barnett's Barn, Broad Chalke).

advantages of the plan were economy in the cost of building and that the crops would probably be drier than in a normal barn. Some writers also thought that a raised boarded floor was better for flail threshing than a solid one. The main defect was access, for which a ladder was necessary. There is a good example by the main road at Washford Farm, Williton, Somerset. A variation with small doors is found in Cornwall, sometimes with fairly low walls and sometimes for machine threshing; it was called a chall barn.

A third variation is the L plan barn (fig. 18b), found in part of Wiltshire. This is a barn with one or two threshing floors which has been bent through 90 degrees, in smaller barns producing a more compact building. This plan also improved the shelter provided for the adjoining foldyard, which was set in the angle. J. C. Loudon noted the danger with a long barn of the wind blowing along its front.

A further, rare variation was the raising of the barn on staddle stones; isolated examples are found in parts of the south of England, of necessity with timber-framed walls. The use of staddles presumably made them drier and more vermin-proof than normal barns. Another possible reason is that the building may have been provided by a tenant farmer; being on staddles, it remained his property at the end of the tenancy.

Plans modified for machinery

Barns where the design was modified by the use of machinery, either for threshing the crops or for preparing feed for farm animals, date, with a very few exceptions, from the nineteenth century. The designs used vary considerably. Because of this, the disappearance of much of the machinery during and after the Second World War, and changes in the use of the various rooms, interpretation is less

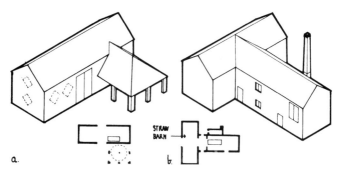

Fig. 19. Barns with fixed threshing machines: a, in a traditional threshing barn, with a horse engine house, based on Gold Hill Farm, Bosbury, Herefordshire; b, with a stationary steam engine, based on a Northumbrian example.

easy. For simplicity, the effects of the different types of machinery will be examined separately.

The first reliable threshing machine was made in 1786 in East Lothian by A. Meikle. Using machinery had a number of advantages over the flail: the work could be done in a few days, rather than spread out over many months, so reducing the need for barn room and losses by damp, rodents and birds in the barn or ricks; the grain was threshed more cleanly, especially if the straw were damp; it was not bruised, as happened with the blows of the flail; and pilfering was prevented or reduced. Whilst flail threshing was a monotonous occupation it did have some advantages: it provided work in winter, was considered more suitable for malting barley and did not break the straw as machine threshing did, making it unsuitable for thatching anything more permanent than a rick.

Threshing machines were initially more used in the north than in the south of England, although they seem also to have been used early in the south-west and in Wales. At first the machinery was fixed, standing within the barn, so that the crops had to be brought to one place for threshing, whether from ricks or from other barns. Some early barns built to house threshing machines were of the old type 1 or 3 plan (fig. 19a). The machine stood end-on to the threshing floor, close to a side wall. Being fairly tall, it required a loft for feeding the corn into it, which covered part or the whole of the bays in that half of the barn. The bays opposite were probably used for threshed straw.

The introduction of machinery did, however, lead in some cases to the complete replanning of the barn. The threshing floor disappeared. One end of the barn was completely lofted, the corn being fed into the building at first-floor level through doors or pitch-holes (fig. 19b). The other part was a straw barn, open to the roof. In some cases there was a similar corn barn at the other end of the lofted section, to feed it, but these seem to be early in date. Some West Country threshing machines were low and did not need a loft to feed them: they stood in a relatively low barn, set at first-floor level over buildings for cattle.

Reliable portable threshing machines became available later than fixed ones and were much more used in the south and east of England than in the north. They could be used in two ways. The machine could stand on the threshing floor of an orthodox flail-threshing barn, which provided shelter from the weather: barns were built for this purpose throughout the nineteenth century. Alternatively, the threshing machine could be taken to the ricks, whether at the farmstead or out in the fields, and thresh in the open. The barn then became redundant: it might survive as a covered rick house or for housing straw. If the farmstead were rebuilt it could well disappear.

The second type of machinery was known as barn machinery, even when it was used elsewhere. It was for preparing feed for the livestock, a practice which largely developed during the first part of the nineteenth century. It was often concentrated at the barn: that was where straw was traditionally kept, it was near the rickyard, and although most of the machinery could be hand-driven it was more efficient when driven mechanically and could there use the

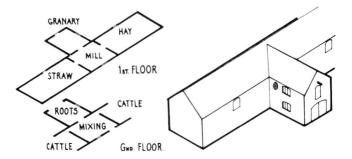

Fig. 20. Barn for feed preparation only: exploded plan and aerial view, with wheel for drive by portable steam engine. Based on a Staffordshire example dated 1863.

same source of power as the threshing machine. Hay and straw were cut up into short lengths, confusingly called chaff. Turnips, swedes and potatoes had to be sliced or pulped, as cattle tended to choke if fed them whole. Linseed cake came in large slabs, which had to be broken up. Oats might be bruised, beans ground and a flour mill might be useful. There was even a short-lived vogue for boiling or steaming some types of feed.

Where there was no barn, for either flail or machine threshing, lofts were usually built for hay and straw, supplying a room or space where the chaff cutter stood. This space might alternatively be supplied by a combination of barn and loft (fig. 20). With the chaff cutter would be the cake breaker and the chute to the hopper on the mill. Chaff and broken cake fell through holes in the floor to the mixing room below. Off this, or as part of it, was the root or turnip house, where these were pulped or sliced. It usually had wide doors, permitting the unloading of a cart by tipping. The actual arrangement of the rooms might vary, but this was the basic process.

A simple arrangement was to build a two-storey section at the end of a traditional barn, sometimes combined with a loft in it. The commonest plan, however, was merely to insert a loft in one end of an orthodox flail-threshing barn, to carry the chaff cutter and cake breaker, with a mixing room below (fig. 21). The latter might also be the turnip house, or that might be a separate room adjoining the cowhouse.

Some provision for feed preparation was found on nearly all farms by the end of the nineteenth century. The lofts or extra rooms usually survive, bereft of their machinery, which has been sold as scrap. An example of the last type can be seen re-created at Acton Scott in Shropshire.

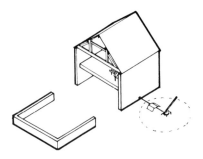

Fig. 21. Barn with loft inserted for feed preparation, with horse engine in open air to drive chaff cutter.

Motive power for threshing and barn machinery

Various sources of motive power were used to drive threshing machines and later barn machinery. Horse, water, wind and steam power were all used from the early part of the nineteenth century, often with distinctive buildings. Gas and oil engines had begun to appear by the end of the century. In each case the source of motive power is separate from the machinery driven.

Early horse engines were housed in a building, the horse engine house, which was one of a variety of geometrical shapes (figs. 19 and 22). It stood against the barn, nearly always on the side away from the farmyard; very occasionally there was a loft or room above. The roof might be supported by piers or pillars or by a wall with windows; the openings were to give the ventilation necessary because of the hard work involved. Early designs needed four to six horses to drive a threshing machine, but later improvements reduced the number; with lighter duties only one or two would be needed. The horses turned a vertical spindle, which in its turn drove a horizontal shaft running overhead into the barn. The distribution of horse engine houses is very uneven: they are common down the east coast from Alnwick to the Humber, in parts of the West Country, Scotland and the former county of Cumberland, but otherwise only scattered examples are known.

An alternative, common in some areas, was to use the horse engine without a shelter, taking advantage of the cast-iron type which had appeared in the 1840s. In these the rotating shaft is at ground level, the horse stepping over it (fig. 21). Occasionally the circular walk can be seen, but otherwise little evidence of this type usually survives.

Steam power was also used early, the first known example on a farm being dated 1798, in North Wales. Initially, the engines were built into the farm buildings, with a prominent factory chimney

Fig. 22. Half-octagonal horse engine house on side of barn.

(fig. 19b). They had the advantage over the horse engine of not tiring if kept well fed. Although costly to install, they were considered cheaper than the horse engine on farms larger than 200-300 acres (81-121 hectares), writers differing as to the minimum size. Examples are scattered throughout Britain.

Portable steam engines began to appear by the early 1830s. They had the great advantage that they could be used to drive machinery anywhere on the farm; some were designed for ploughing. They were used much more than the fixed steam engine and were fairly widespread, at least in the Midlands and south of England, by 1854. They could stand outside the barn, with belts to drive a wheel on the end of shafting (fig. 20). Sometimes long, high porches were built to shelter them when driving a threshing machine in the barn, possibly with a cast-iron flue in the roof to take the smoke and sparks.

Water power was sometimes used in hilly areas but could be expensive to install. The wheel was normally outside the barn, under a lean-to extension; very occasionally a turbine was used. Wind power was even more expensive to provide and was difficult to control, so it was rarely used. Some ten examples survived in Scotland in 1979 and there is, or was, one just outside Brighton. Wind or water mills for grinding corn are excluded from this study, as they are not farm buildings, although a farm might sometimes be attached to a mill.

The hay or Dutch barn

The hay or Dutch barn is completely open on one or more sides, these consisting of brick or stone piers, or of pillars to support the roof (fig. 23). They could vary considerably in size, from one to ten

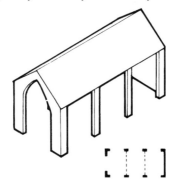

Fig. 23. Dutch barn, based on Cheshire examples.

or more bays, and were sometimes attached to other buildings. Some, principally in Cumberland, were built over accommodation for livestock. They were built to house hay or straw, or possibly unthreshed corn, and had the advantage over ricks that thatching was saved and some immediate protection was provided for the harvest in wet weather. Apart from in a few areas, they were very rarely built before the late nineteenth century, being considered an extravagance. The exceptions seem to have been Cheshire, south Lancashire, parts of North Wales and Cumberland, all built largely for hay. Their use increased considerably after 1885, however, when they were built by landowners to combat the depression. These later ones were generally iron-framed, with corrugated-iron roofs.

3. The cartshed

Before the nineteenth century farmers' horse-drawn equipment consisted largely of carts, waggons, ploughs and harrows. More complex and specialised equipment appeared thereafter, often used only seasonally: the more common were drills, threshing and reaping machines and others used for cultivation, like subsoil ploughs. Shelter was necessary, even for the simpler items, for those made of wood suffered from the sun and the rain, whilst the more complex, with moving parts of iron, were liable to rust and seize up if left outside.

Cartsheds were probably found only on some larger farms before the eighteenth century: where there was but little equipment it could be kept on the threshing floor, in the porch attached to the barn or in a covered way (fig. 17). One writer noted that ploughs were rarely housed, being in regular use, but he was probably thinking of iron ones. Other equipment might be left outside covered with a little straw as a temporary thatch. In some parts cartsheds were unnecessary, even up to the late eighteenth century: for example in south Devon farm carriage was still by packhorse until after 1770.

The most common type of cartshed found surviving on farms is an open-fronted building, approached from the side, like open sheds for yard cattle or horses (fig. 24a, b). The openings were separated by brick or stone piers, timber posts or cast-iron columns, with a shallow arch or timber lintel above. The supports, which varied in size depending on the material used, were generally spaced to give an 8-10 foot (2.4-3.0 metre) opening; internally the building would be between 16 and 22 feet (4.9 and 6.7 metres) deep, but there were some exceptions, usually larger. The cartshed differed from buildings for cattle or horses in that it rarely opened on to the foldyard. It might simply turn its back on the yard,

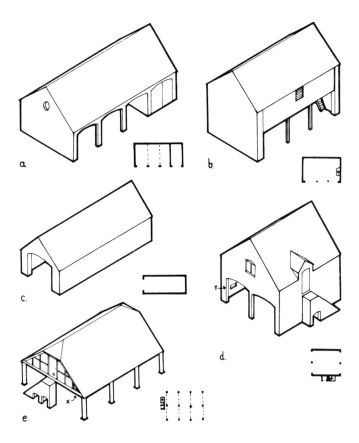

Fig. 24. Cartsheds, b, d and e with granaries over: a, entry one side, with one doored bay; b, entry one side; c, long, end entry; d, entry on two opposite sides, with cupboard (y); e, fully open, with ladder rack (x) and cat opening in granary door (now at Avoncroft Museum, Worcestershire).

1. *Farmstead with multiple yards, c.1860; Little Langford, Wiltshire.*

2. *Decorative farmstead, 1858; Eastwood Manor, East Harptree, Somerset.*

3. *Cartshed and granary disguised as church, mid eighteenth century, with seventeenth-century barn on left; Seighford Hall, Staffordshire.*

4. *Type 1 barn, c.1700, with type 3 cowhouse and granary over on far end: note the high brick plinth; Gilbert End Farm, Hanley Castle, Worcestershire.*

5. Type 1 barn; Rectory Farm, Longdon, Worcestershire.

6. Type 2 barn with lean-to shelter sheds: far end seventeenth century, extended towards camera eighteenth century; Sheepcote Hall, near Stowmarket, Suffolk.

7. *Type 3 barn with low, wide doors and pitch-holes, nineteenth century; Colston Bassett, Nottinghamshire.*

8. *Type 5 barn, with sliding doors and cowhouse in near end, mid nineteenth century; Kilnside, Bishopsdale, North Yorkshire.*

9. *Bank barn (type 1), over sheds for loose cattle; nineteenth century; Cothelstone, Somerset.*

10. *Steam engine and rooms for feed preparation, 1864; Preston Vale, Penkridge, Staffordshire.*

11. *Horse engine house, mid nineteenth century; Bury Barton, Lapford, Devon.*

12. *Cartshed with granary over, mid nineteenth century: note decorative gable; Upper Mitton, Penkridge, Staffordshire.*

13. *Cartshed with granary over, nineteenth century; Craster South Farm, Embleton, Northumberland.*

14. *Cartshed with one doored section, 1850; Red House, Church Eaton, Staffordshire.*

15. *Cartshed with timber posts, after 1877; Manor Farm, Cogges, Oxfordshire.*

16. *Granary over cider house, nineteenth century; Manor Farm, Middle Littleton, Worcestershire.*

17. *Stables, type 2, far set 1819, near after 1862; Perreton Farm, Arreton, Isle of Wight.*

18. Stables, type 2, with buildings for cattle beyond, nineteenth century; farm near Padstow, Cornwall.

19. Stables, type 2, c.1840, with later stable and barn beyond; Red House, Church Eaton, Staffordshire.

20. *Cowhouse, type 1, with feeding passage, mid to late nineteenth century; Place Farm, Tisbury, Wiltshire.*

21. *Worcestershire cowhouse (type 1), without a front wall, nineteenth century; Church End Farm, Twyning, Gloucestershire.*

22. Field cowhouse, latter half of the eighteenth or early nineteenth century; Newbiggen, Bishopsdale, North Yorkshire.

23. Type 4 cowhouse, nineteenth century; Robert Hall, Tatham, near Lancaster, Lancashire.

24. *Two type 5 cowhouses, in range attached to the house; nineteenth century; Shielhill, Kirkwhelpington, Northumberland.*

25. *Shelter sheds, for loose yard cattle, nineteenth century; Hill End, Castlemorton, Worcestershire.*

26. (Above) A linhay, mid nineteenth century, a shelter shed with open loft above; Bury Barton, Lapford, Devon.

27. (Left) Covered yard, c.1914, being roofing of open yard surrounded by cowhouses; Manor Farm, Bradmore, Nottinghamshire.

28. *Looseboxes, second quarter of nineteenth century, some doors have been altered to windows; Hatherton, Staffordshire.*

29. *Two pigsties, with chutes to fill troughs, second half of nineteenth century; Coverside Farm, Hanley Castle, Worcestershire.*

30. Dovecote, consisting of nesting boxes on the gable, 1758; Village Farm, Seighford, Staffordshire.

31. Dovecote, thirteenth or fourteenth century; Sibthorpe, Nottinghamshire.

opening on the other side; alternatively it opened on to the drive or, occasionally, faced a yard of its own. These positions were used to prevent livestock getting into the building and damaging themselves or the equipment; it also prevented them accidentally getting out of the yard when the gate was opened to take out a cart or other piece of equipment. There was still the problem of cattle being driven past on the way to the fields, so occasionally gates were fitted across the openings.

Some cartsheds or parts of a larger range had full doors (fig. 24a); these were provided to give better protection from the weather and sometimes also from malicious damage to the more complicated equipment. Those with high doors were probably for portable threshing machines or steam engines. Where the doored section formed part of a larger range it was separated from the rest by a solid wall. The shelter for the farmer's trap or gig had doors, looking like a single-opening cartshed. It might form part of the cartshed range but was more usually separate, with its own stable nearby.

Internally, the cartshed was open to the back, but the bays might be partly divided, either by extra posts or by short spur walls carrying the roof or the floor above. A characteristic feature is a cupboard-like recess found in a side wall near the entrance, intended for a grease horn and small tools (fig. 24d). A good example can be seen at Manor Farm, Cogges, Oxfordshire. A few cartsheds have two or more of these.

The size of a cartshed could vary considerably, from a single bay to a long range; on larger farms there might be more than one set of buildings. The commonest plan to survive, as has been noted, is that with openings only on one side, usually the long one. This arrangement provided reasonable shelter for the implements and had the great advantage that the contents of each bay were readily accessible. Most writers recommended that they should face north to avoid the sun, but this advice was by no means always followed. In a few cases an end bay of the range was walled in on the front instead of having an opening: this section was probably for ploughs and harrows, or to give better shelter to a reaping machine or drill.

Some cartsheds were entered from the end, that is under the gable. They are somewhat deeper than normal (fig. 24c). This type is fairly common in Worcestershire. A few were very much deeper, the largest that the author has seen extending 51 feet (15.5 metres) back from the entrance, whilst being only 11 feet (3.35 metres) wide. It must have been very awkward to use, time being wasted in getting equipment out from the back.

A much less common alternative plan was to have the cartshed with openings on opposite sides or ends (fig. 24d). It permitted easy access to equipment housed back to back and also enabled a waggon to be pulled in and later out without reversing. This would be particularly useful if there were a granary above, with provision for lowering sacks of grain through the floor.

A less normal type to survive is open on two or more adjacent sides, sometimes all round, like the example re-erected at Avoncroft, Worcestershire, which has circular brick piers (fig. 24e). Whilst this type was economical to build and provided good ventilation for the equipment, it provided poor shelter from driving rain and snow, which could be blown right through.

A small room with a normal-sized door might be provided opening off or adjoining the cartshed: this was for paint, spare parts and tools used to repair the implements. It might also contain some hand tools, such as shovels or pitchforks. Its provision seems to have been restricted to larger farms. Hand tools might otherwise be leant against or hung on the back or side walls of the cartshed; sometimes they were stored in a loft made in the roof over part of the otherwise single-storey building.

Other such triangular lofts might be used for hens. More often, if there were rooms or a loft over the cartshed, they had low or normal-height side walls: the commonest use was as a granary. Where there was no loft, or where it did not cover the full width of the building, ladders might be stored resting on the tie-beams of the roof trusses: they were fed in from the end, an opening being left for the purpose (fig. 24e).

4. Other buildings for crops

The granary

The granary was a room or building used to house the grain after it had been threshed or winnowed, whether it was for sale, for use as seed or for feeding the livestock. Its size depended on a variety of factors: of these the acreage used for grain crops when it was built is perhaps the most important. Speed of threshing and the length of time that the grain was to be kept were other factors. Some farms had more than one granary.

Whether it was a building on its own or a room within a larger building, the floor of the granary was nearly always raised above the ground; it might be raised only 2 feet (600 mm), or it might be enough to make it a first-floor room. This was to keep it dry and to provide some protection from vermin. First-floor rooms were often approached by an external staircase. Where this was of brick or stone the top landing was invariably set one step down from the granary door, which came forward to the edge of the step. This was probably to provide some additional protection to the bottom of the door, making it harder for rats to gnaw through. It would also prevent any water draining off the landing into the granary. A recess was often formed under the stairs for a dog kennel, and a hole was sometimes left in the granary door for the cat (fig. 24e)!

Dealing initially with granaries that are at first-floor level, one of the most recognisable places is over a cartshed (fig. 24); fresh air circulating underneath with the partially or wholly open building made this a dry place. Some of the larger examples present a fine appearance, with rows of arches below matched by equally spaced windows above. In some areas it was normal to put the granary over the stable, as north and west of Ripon, and occasionally in Worcestershire. A number were built over cowhouses or other buildings for cattle, although grain was said not to keep well in such a position. A nineteenth-century development was to make the granary part of the feed preparation rooms. One interesting regional variation is found in the Cotswolds, where the granary was occasionally built over a porch to the barn. Some survive attached to the farmhouse, usually now approached by an external stair and without any access to the rest of the dwelling. This is a survival of earlier practice, as, when upper floors first appeared in the sixteenth and seventeenth centuries in the normal farmhouse (as opposed to large ones and manor houses), they were used for storage, not for sleeping. They were then, however, approached from inside the house.

Sometimes the granary was a separate building, usually square and standing on staddles (fig. 25a, b). Its walls were timber-framed, clad in weatherboarding, generally with a pyramidal or gabled roof. This type is found in southern England. A much rarer alternative, shown in fig. 25c, was to build it of brick or stone, the walls rising from arches set just above ground level. The approach to both was by timber steps. There might be more than one floor, access to the upper ones being by internal stairs or ladder.

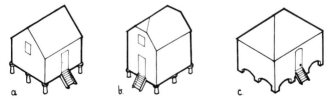

Fig. 25. Granaries: a, on staddles (at Droxford, Hampshire); b, on staddles, two floors (at Bentworth, Hampshire); c, brick (at Dunster, Somerset).

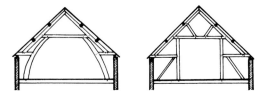

Fig. 26. Examples of trusses in low-walled lofts.

The windows were generally needed more for ventilation than for light: they might have fixed or adjustable louvres (fig. 24b), or sometimes shutters, with metal bars. The internal features depended partly on how the grain was stored, partly on the type of roof trusses. The grain might be stored loose or in sacks in a completely open room, or in bins, in which case the room was divided by low, wooden partitions. The type of roof truss was affected by the height of the side walls (fig. 26). Where these were low it was necessary to have a truss without a continuous tie-beam to permit access throughout the granary. In a few, brick spur walls were used, as in barns. Studying such trusses is an interesting aspect of studying farm buildings.

The walls might be plastered, and a plaster ceiling was sometimes provided: these presumably were for cleanliness and to reduce the amount of dust-collecting surfaces. Such ceilings have often been removed. The only other upper-floor farm building to be plastered was accommodation for farm labourers. In some areas the floor might be of plaster, rather than the normal boarding. This may have been to provide a better surface for shovelling grain, if it were kept loose: it was also less likely to be holed by vermin.

When the granary was over a cartshed, a trapdoor could be provided in the floor for loading a cart beneath: the wheel for the rope may still survive in the roof. A floor over a cartshed could readily be strengthened to support the weight of grain by introducing pillars or posts across the building or by building spur walls. Such extra provision was not easy with any other type of building beneath.

The hopkiln

Hopkilns were used for drying the hops before they could be despatched to market. They are found in the hop-growing areas of Kent and east Sussex, where they are called oasthouses, the Farnham area of Surrey, Herefordshire and west Worcestershire. Hops were grown before the nineteenth century in a few other areas; they were first grown in England in the sixteenth century.

The most obvious kilns surviving are the square or circular nineteenth-century type, which are tall rooms, with a characteristic tall, conical roof, terminating in a cowl (fig. 27). They were attached to a normal two-storey building. Circular kilns first appeared in Kent about 1805, soon becoming the normal pattern there, but were not used in the Farnham and Herefordshire areas until after 1835. Square kilns were again being built in the Herefordshire area by the 1870s but were not used in the southern areas until shortly before the First World War.

The kilns had brick walls: inside there was an open brick box to contain the fire, the draught being controlled by the door. In some cases there may have been more than one fire, each in an iron box. The drying floor was some 10-11 feet (2.16-2.47 metres) above, made of laths or battens set fairly close together, the hot air from the fire passing up between them. The hops were spread on a horsehair cloth laid on the battens, access being by a door in the side of the kiln. Above was a steep-sided cone, usually timber-framed like a roof, with a plastered internal finish and slates or tiles externally. A very few cones were of brick.

An alternative, occasionally found in the south, was to have an enclosed fire, with a series of brick or iron flues to heat the air, leading to a chimney. This could burn any type of coal without flavouring the hops: an open fire was limited to hardwood, charcoal, coke or later, with the coming of the railways, Welsh coal. After 1885 a few rectangular kilns with a steep gable roof and a ventilator

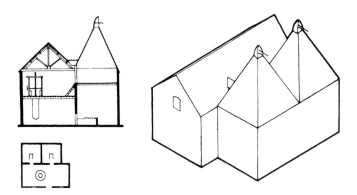

Fig. 27. Hopkilns, Herefordshire type. Section shows press for hop pocket.

running the length of the ridge appeared in the Herefordshire area.

Adjoining the kiln was a two-storey building: the upper floor was some 3 feet (0.9 metres) below the drying floor, which was reached by steps. It was used for receiving the hops and for spreading them out to cool after they had been removed from the kiln. They were then put into pockets; these were long sacks which were suspended through a circular hole in the floor, called the treading hole (fig. 27). Until presses were introduced the hops were compressed by a man treading them in.

The lower floor was only required for stoking the fires and to give space for the pockets to hang. Otherwise it had no connection with hops and was used for cidermaking, livestock or as a cartshed.

Earlier kilns, before the first part of the nineteenth century, were square. They were built into a normal, gabled building. The fire was in an open brick box as in the later ones, but above was a brick or lath-and-plaster cone up to the drying floor. This was covered by a normal roof, the smoke escaping through a square ridge ventilator or through ridge tiles set at right angles to the line of the roof. Such kilns appeared in the seventeenth century and might be in a separate building or in a wing of the farmhouse, as was usual in Herefordshire. An alternative was to spread the hops out to dry, without any artificial heat, in the parlour or a bedroom, a process taking two or three weeks. This early, secondary use of these rooms is recognisable by the circular hole in the floor for the hop pocket.

The maltkiln was not dissimilar; the main differences were that all floors of the adjoining building were used for malting, the kiln was always square and the drying floor was covered with tiles with small holes punched in them. It is sometimes found on a farm but is more often a separate, industrial building and so is not described here.

The cider house

The cider house, which first appeared in the seventeenth century, was the room where apples were crushed and pressed for their juice, which was then put into barrels to ferment. It might be either a separate, single-storey building or the lower floor of a larger one. In Herefordshire and Worcestershire it might be combined with the hop rooms, occupying the otherwise superfluous ground floor. Alternatively, there might be a granary over, as at Middle Littleton Tithe Barn, Worcestershire, or a loft for storing apples, as was the practice in Devon. The apples were otherwise kept in heaps outside.

The cider house contained two characteristic pieces of equipment, shown in fig. 28. The first was the apple mill, consisting of a circular stone trough with a stone roller which crushed the apples.

The roller was on a wooden arm, connected to a revolving, vertical spindle in the centre of the trough; it was driven by a horse walking round. The second item of equipment was the press, two vertical members rising from a stone base with a beam operated by a screw or levers. A timber, or stone, tray with a rim was put on the base to catch the juice and direct it to a tub. The crushed apple was placed on the tray in 4 inch (102 mm) layers, separated by cloths. The juice was transferred from the tub to a barrel to ferment. The crushed apple might then be reground and repressed. Most presses had heavy timber frames but some later ones were of iron. The residue apple might be fed to the pigs.

The barrels for fermenting the juice were normally kept in the cider cellar, which was not necessarily connected to the cider house. It might be a cellar under part of the farmhouse, a lean-to on the back of it, or some other separate room, probably with a floor partly below ground level. Once the juice had fermented enough, it was transferred to a fresh barrel for storage.

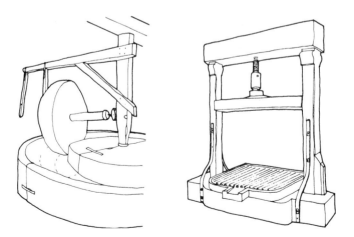

Fig. 28. Cider mill and press, Worcestershire examples.

5. The stable

Two types of horse were found on the farm: the waggon horse, used for agricultural work, and the hackney, used for riding or pulling the carriage or trap. On larger farms they had separate stables, which might be built adjoining one another. The principles governing both were similar, but we are here concerned with the waggon stable, of which there might be more than one.

The relationship between the amount of stable accommodation and the size of the farm was affected by a number of factors. Perhaps the most important was the use of oxen as draught animals: their use for some farm work continued in parts of the Cotswolds and Sussex into the early years of the twentieth century but had otherwise largely died out by the late eighteenth century, if not before. In some areas horses had been used for all farm work from the Middle Ages and most medieval farmers appear to have had at least one. A second factor was the amount of arable land on the farm, and whether this was light or heavy: if the latter, more horses would be required to make up the plough team. The position of the farm buildings in relation to the arable land of the farm and the compactness or otherwise of the holding also affected the number of horses by dictating the length of journey necessary before work could begin. Thus the enclosure of open fields and the regrouping of fields in enclosed areas to form compact holdings during the eighteenth and early nineteenth centuries reduced the number of horses needed. However, greater numbers became necessary again with the increase in farm work during the mid nineteenth century, including the use of horse engines for driving farm machinery. Where farms had been too small to support a full plough team, communal ploughing seems to have been adopted, the team being made up by horses from more than one farm.

Now that horses are no longer used for farm work, the stables may be left empty or used for calves, the fittings being unaffected. More often they have been partially or wholly gutted for conversion to some other use, usually cowhousing. The stable, however, differed from the cowhouse in a number of ways, some of which may still be apparent.

Beginning with the exterior, before the mid nineteenth century the stable was much more likely than the cowhouse to have a window; this might be closed by a shutter or louvres rather than glass. Alternatively there might be ventilation slits. Where both buildings had windows, those in the stable tended to be the larger as the importance of light and ventilation had been appreciated earlier for that building than for the cowhouse. The door to the former might well be wider and sometimes a horseshoe was fixed above it. Most stables had lofts above them; only a small proportion were single-storey, these being built after the third quarter of the eighteenth century. In contrast loftless cowhouses appeared in the early nineteenth century and became usual in many parts of Britain.

Internal features

The loft was nearly always set higher in a stable than in a cowhouse; with a ceiling height of some 8 feet (2.45 metres) or higher, it gave more air. Although opposed by some writers on grounds of ventilation, the loft survived for its insulation value: it helped to keep in some residual heat during the day whilst the horses were out working, and enabled it to warm up more quickly when they returned. Where there was no loft the insulation might in part be supplied by inserting a plaster ceiling. Lofts were also useful for storing hay, which could readily be fed into the racks below. Access was by a vertical ladder from inside the stable. Not all lofts over stables were used for hay or straw, however.

The most obvious fittings were the partitions between the stalls (fig. 29a). There was usually one for each horse, but occasionally horses were kept in pairs, with a single stall at the end in case there was one animal given to kicking. These partitions were of wood, high at the head and normally the full length of the stall, to prevent the horses biting or kicking each other. They were set wide apart, some 5 or 6 feet (1.5-1.8 metres), so that there was enough room for the horse to be groomed whilst tied up. Stalls were smaller in a cowhouse and set for two animals, each allowed about 3 feet. As in the cowhouse, there was a channel at the back of the stall, to which the floor sloped gently.

At the head of the stall were the trough and rack for feeding, with a tethering ring in the front of the former. The trough was high, about 3 feet to 3 feet 6 inches (0.91-1.07 metres) at the lip, as the horse fed standing. It was made of wood, brick or stone; the last two types might have a timber lip, which was better for the horse's mouth. Above there was nearly always a wooden rack, the front sloping out from the wall; it stretched the full width of the stall. Bowl-shaped cast-iron ones are a late development and, like cast-iron troughs, are usually found only in hackney stables. On most farms, however, the hackney stable had the same type of fittings as the waggon stable. In a few cases the front of the rack was vertical, thus reducing the danger of seeds falling into the horses' eyes; this was an expensive luxury as, unless there was a passage along the head of the stalls, it required a wider building. Feeding passages are very rarely found in stables; as the stalls were wide, the feed could easily be carried past the horses without the risk of it being spilt.

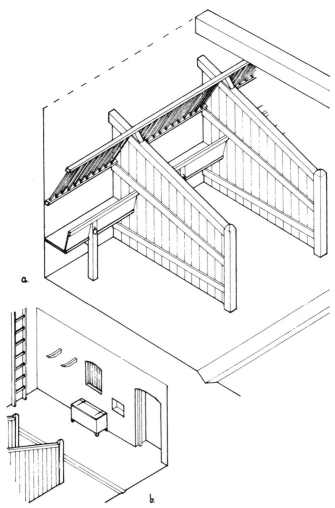

Fig. 29. The stable interior: a, showing stalls, trough and rack; b, looking towards door, with cupboard, corn chest, harness hooks and ladder to loft.

Behind the horses, near to the door, there might be one or more cupboard-like recesses (fig. 29b), which were for candles, curry combs and medicines. Where such recesses were not feasible wooden cupboards were used. A number of wooden hooks may project from the walls: they were for hanging the harness, one for each horse. Sometimes grouped at one end, in other stables they approximated as nearly as openings permitted to a position behind the stalls they served. In a number, however, a separate harness room was provided, or part of the stable partitioned off for the purpose (fig. 30e). This room had, of course, no provision for tying or feeding horses, but had a door to the stable. Separate provision was sometimes made only for the hackney stable.

A separate room might be provided for the horses' feed, but this is relatively rare. Sometimes it also served the cowhouse and on occasion it was combined with or opened off the harness room. It was particularly useful for hay or straw if there were no loft over the stable or if the loft were used for some other purpose. When chaff, in this case the husks of threshed grain, formed an important ingredient in the feed, as in Suffolk, a separate room was provided to house it or a corner of the stable was enclosed for the purpose: in both cases the bottom of the access would have removable boards to retain the contents. Grain for the horses was kept in a wooden chest, which sometimes still survives. This could stand in the feed room but otherwise normally stood against the back wall of the stable.

The plan

Stables were planned in four different ways. In that which was used in most of the surviving early stables, the horses were tied facing along the building, parallel to the ridge of the roof (fig. 30a). Where there were only three or at most four horses, this was the most economical arrangement, the building often being attached to the end of a larger range. Where more horses were to be housed the plan could be doubled, the two rows facing away from each other, with a single manure passage (fig. 30b). A contract for building a stable in Suffolk, dated 1473, describes this plan. Only rarely was it turned round so that the horses faced each other over a feeding passage: as has been noted, the passage was not very necessary and required a larger building. There were a few rare cases where the horses faced a row of cows across a feeding passage.

The second plan, with the horses facing across the building, was much more flexible, being adaptable for any number of horses from one upwards, unlike the first plan, which worked in multiples of three or four (fig. 30c-f). Even with a large number of horses all could be in one stable, which had advantages for supervision and

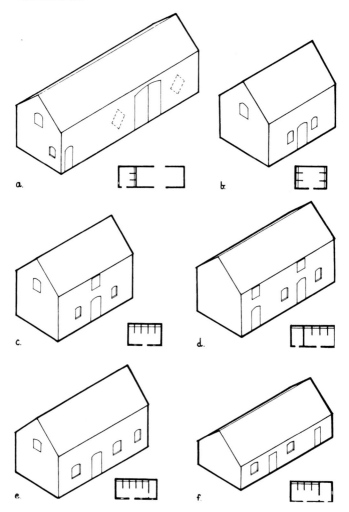

Fig. 30. Stables: a, type 1, attached to a barn; b, type 1, with two rows of horses; c, type 2; d, type 2, with loosebox adjoining; e, type 2, with harness room adjoining (same plan might be used for a feed room); f, type 2, with feed room adjoining.

issuing orders. Some, however, were still divided into two. Better ventilation for each horse was possible. Most separate harness and feed rooms relate to this arrangement. It became the normal plan throughout Britain in the nineteenth century, although the period when it displaced the first varied from one part to another: in western Staffordshire, for example, this took place in about 1800, but not until forty years later in north-east Wales. Some earlier examples survive, however, dating back to the early sixteenth century.

The third type of stable is the loosebox, in which the horse was untied and free to move about: it was thus larger than a stall. One or more separate rooms might be provided adjoining the stable, or part of it was partitioned off to form a box. They were more necessary for hackney than for waggon horses, which were taken out every day to work. They were also used as foaling pens, to house the foal during the day whilst its mother was working, or for a sick horse. A separate room was more important in the last case.

The fourth type of stable was found in Suffolk and some other parts of the country in the early nineteenth century; in some areas it continued to be built into the twentieth century. It consisted of a normal stable, opening off a yard with open shelter sheds, the horses being turned out into the yard at night. As the stable was only used for grooming and feeding, there might be few or no partitions. There were racks and troughs in the shelter sheds for extra feed. Horses housed in this manner were said to be less liable to disease.

6. Buildings for cattle

Cattle were kept on all farms, but the reasons for keeping them varied. One of the most obvious reasons today is for dairy produce: however, until the late nineteenth century butter and cheese were much more important than liquid milk, except perhaps in the vicinity of some large towns. Breeding of calves appears to be related to this, but the two did not always go together, as some breeding areas had little market for dairy produce, and the calves needed milk. A third reason is the production of meat, with leather as a by-product. There are, however, two other important but less obvious reasons. The first was producing manure for the crops, mainly by converting hay and straw; in some cases this was the principal reason for keeping cattle. The other was for ploughing and carting; in many areas cattle were so used until the eighteenth century. In parts of Gloucestershire and Sussex this practice continued until after 1900.

The number of cattle kept depended partly on the type of farming practised; it was strictly limited before the nineteenth century by the amount of feed grown on the farm, in particular the amount available for the winter months. The idea of a large scale slaughter in the late autumn is now completely discredited. The number of cattle in Britain had been increasing during the eighteenth century and was to increase even more during the nineteenth, especially between 1864 and 1874, when numbers rose by one third. This growth was encouraged by an increase in the demand for livestock products, and because the prices paid for them did better after 1815 than those for corn. It was made possible by the adoption of crops specifically grown to supply cattle feed, which also helped to improve the fertility of the soil, and later by the use of purchased feed.

Most surviving buildings for cattle seem to date from the nineteenth century. This is partly because of the great increase in numbers kept, partly because of a greater appreciation of the benefits of shelter and of improved standards in the buildings, which resulted in the replacement of earlier ones. A number of earlier, post-medieval examples survive, and only a few medieval ones, much fewer than for barns. Unlike barns, they are small in size, although there do appear, from documentary sources, to have been some large medieval cowhouses.

The amount and type of shelter provided for cattle varied considerably from one part of England and Wales to another up to the mid nineteenth century. In the sixteenth century it was common to keep cattle in the open over the winter, but there are some references to housing them. In some areas it was still rare to house them at the end of the eighteenth century, for example in Gloucestershire, Worcestershire and Lincolnshire. Housing milking cattle was, however, otherwise then becoming the normal practice. Adequate shelter for all cattle, milking and fatstock, seems only to have become general throughout the country after the mid nineteenth century.

The cowhouse

The cowhouse is the most important of the various types of building used to accommodate cattle. It is a building in which the cattle are kept tied all the time that they are in it; there was little or no difference between those built for milking cattle, fatstock or draught oxen. In certain parts they may be called shippons, byres or cow-stables, but these are purely local terms, restricted to relatively small areas.

The cowhouse was nearly always approached from the main yard in the farmstead. This was for ease in disposing of the manure, which was collected there until it was taken out to spread on the fields. If there were a number of yards, the one in front of the cowhouse was used for its manure, and possibly for loose cattle as well. A few cowhouses had windows at the back for pitching the manure directly into the adjoining field, but it was less wasteful to have one central heap.

Most early cowhouses were built with lofts above: this practice continued in some areas, such as Devon, Cheshire and the Lake District, into the second half of the nineteenth century. Most cowhouses, however, from the early years of that century were built single-storey, unlike stables.

Many cowhouses are now well windowed to meet government regulations; even where the openings are not modern, they generally

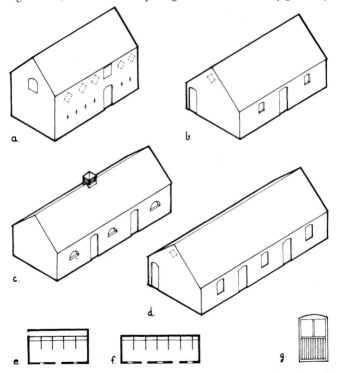

Fig. 31. The cowhouse: a, with loft and ventilation slits; b, with feeding passage and small windows; c, with ridge ventilator and semicircular windows, no feeding passage; d, with feeding passage and larger windows; e, plan of b; f, plan of c; g, window with glass above and hit-and-miss ventilator below.

contain modern frames. This often makes determining the original form of the windows difficult, but fortunately in most areas a number of cowhouses have survived unaltered. Early cowhouses, with few exceptions, had no windows, although they might have slits for ventilation. As the buildings were generally small, enough light for cleaning or inspection could be obtained from the door. Early windows were small or semicircular, closed by shutters. Larger-sized openings could be filled with hit-and-miss ventilators or louvres, in some cases with fixed glazing above (fig. 31g). The need for light in cowhouses was becoming generally recognised by the mid nineteenth century. In some areas, however, windowless cowhouses were still being built up to the end of the century: if they were single-storey, light might have been provided by areas of glass slates or tiles on the roof.

Alternative sources of ventilation were holes or slits, patterned as in a barn: occasionally ridge ventilating tiles were used and, very rarely, a louvre (fig. 31c).

Internal features

The first internal feature to be noted is the roof or floor above. Early cowhouses had low ceilings with a loft over, lower than for a stable; later they became higher, as cows were found to fall ill with too little air. A large proportion of cowhouses built after 1800 were open to the roof. The main objections to lofts arose from the cows being kept tied all the time, the problem being to dispose of the heat, not to conserve it, as in a stable. There was also a danger of dust falling on the animals from the floor above. Some lofts have been removed to meet government regulations.

The fittings for tying and feeding the cattle have generally been replaced since 1945 to meet government regulations; some earlier ones, however, survive in cowhouses to which they did not apply (fig. 32). There was some variety in the arrangement of the traditional fittings, depending on the area of the country and sometimes on their age. The most usual arrangement was to have a partition between every two cows: it was lower and shorter than in the stable, with a straight, slightly inclined top; in some cases it was very short (fig. 32b). The partitions were usually of timber, but in some districts might be stone or slate slabs. The double stall was occasionally subdivided, either by a central post or by a short partition half the length of the main one. In all these the cattle were tied to a post fixed to the side of the partition. An alternative was to omit the partition altogether, cattle being tied to rings in the edge of the trough, or to posts about 3 feet (910 mm) apart which rose to a beam at ceiling level.

The trough ran across the head of the stall; it was usually low,

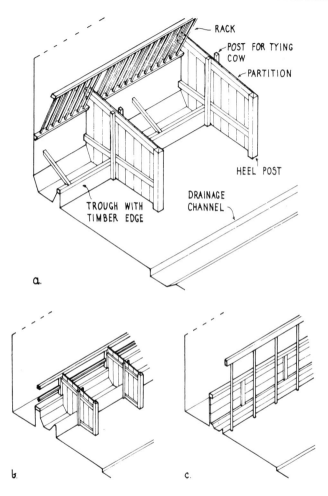

Fig. 32. The cowhouse interior; a, with normal partitions for two animals; b, with feeding passage, short partitions and no rack; c, with feeding passage and no partitions.

unlike those in the stable, ranging from a few inches to about 2 feet (0.6 metres) high to the lip, so that the cattle could feed lying down. It could be made of wood, stone or brick. In parts of the Pennines and Lake District it might be omitted altogether and movable wooden boxes used. Water troughs had begun to be provided for tied cattle by 1845, but very few early fittings now survive: the galvanised iron bowls with a flap operated by the cow's nose are all modern.

Above the trough there might be a rack for hay or straw, nearly always of wood. Unless there was a feeding passage beyond, the front sloped outwards into the cowhouse. Provision of racks seems to have been normal before about 1840 but thereafter they were often omitted. Hay and straw were by then being cut up as chaff to feed to the cattle and were put into the trough; the rack thus became superfluous. Where it has been removed the ends of the top and bottom rails may have left marks in the end walls or on the beams

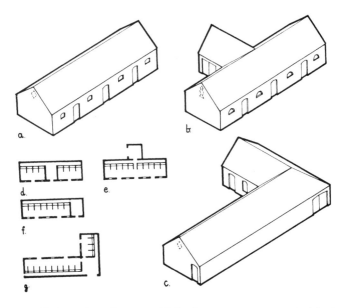

Fig. 33. Type 1 cowhouses, with feeding passages: a, central feed-preparation room; b, projecting feed room; c, L plan, with feed room at corner; d, plan of a; e, plan of b; f, feed room at end; g, plan of c.

above. A variation has been noted in east Sussex: where the cowhouse adjoined a timber-framed barn the front of the rack was part of the barn wall; the back was a net fixed inside the barn.

A passage at the head of the cows is a further feature found in some cowhouses. Its presence is partly dependent on the plan of the building (figs. 33 and 34): it is implicit in types 4 and 6, but only rarely found with 2 and 5; in 1 and 3 it is more likely to be found after 1800 than before. The passage was 3-4 feet (0.91-1.22 metres) wide and connected with the feed preparation rooms or the open air, the feed being then brought across the yard from a separate building. There might be a way through from the passage between the cows to the area behind them (fig. 33e). Some passages are much wider, being used for feed storage and preparation. The most obvious type is the barn floor, appearing in the seventeenth century. Otherwise the wider type seems to be a late-eighteenth-century development, used for housing and preparing turnips (fig. 34d).

The passage was useful in speeding up the feeding of the cattle and in reducing waste of feed: it did, however, require a slightly larger building. If the cows were fed from behind there was a danger of the feed being spilt as it was pushed past the animals, or of it being fouled by their dung. The ultimate in speeding up feeding and in reducing labour was to install a tramway along the feeding passage: small waggons were used, pushed by hand, with either curves or turntables at corners. Turntables were generally used at junctions.

The floor of the cowhouse was paved with cobbles or bricks, and it sloped back gently to an open channel behind the animals, draining to the yard. Some looseboxes or calf pens may have had a slatted floor so that the dung dropped through to a space beneath, but very few survive.

The plan

There are six different ways of arranging the animals, if feeding passages and other rooms opening off the cowhouse are ignored. There were some regional variations in the types used, but these tended to disappear during the nineteenth century. In the commonest type, the cows are in a single row, facing across the building (type 1, fig. 33). It became the normal plan throughout much of England and Wales in the nineteenth century, but was much older in origin, being used from at least the thirteenth century.

It became popular because of the advantages it possessed over most other arrangements. It could be built to accommodate any number of cows from three upwards, whereas types 3, 4 and 5 were limited to multiples of five or six. The cows could be in one room,

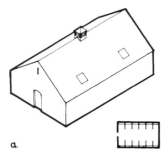

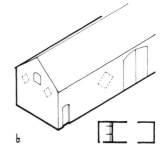

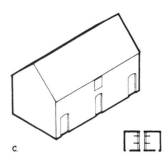

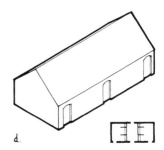

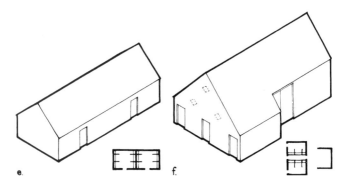

Fig. 34. Cowhouses: a, type 2, single storey, with ridge ventilator and roof lights; b, type 3, attached to barn; c, type 4; d, type 4, with wide feeding passage; e, type 5; f, type 6, barn extending over cowhouse.

68

giving continuous access under cover for feeding or cleaning out; ventilation could be better arranged than in most other types, and the spread of disease was less likely than in types 4 and 6, where the cattle breathed on each other across the feeding passage. Long cowhouses of type 1 might be subdivided into a series of rooms or broken up by feed preparation rooms. Some were L shaped to enclose two sides of a yard. One or more feed preparation rooms might be built connected to the cowhouse, at one end or part way along: in the latter case they may project at the back. Where there was a feeding passage, they always had direct access to it, but they did not always have a door to the area behind the cows.

A variation on this type is the Worcestershire cowhouse, fairly widespread in that county, and occasionally found elsewhere; this had tying and partitions for cattle but was completely open at the front, without a wall (plate 21). It had appeared by about 1800 in this form: one of the cowhouses at Cogges, Oxfordshire, was so built. Later examples may have had a series of part-boarded front walls and doors, and most have since been altered to the more orthodox pattern. Nearly all examples are single-storey. The Devonshire linhay is a separate phenomenon; it is an open-fronted building with an open loft above. While most were used for loose yard cattle, some are reported to have been for tied animals.

The second plan (type 2, fig. 34a) is really the first doubled, with two rows of cattle facing away from each other. Not many survive in use, although it is as old as type 1.

Type 3 (fig. 34b) was invariably built as part of a longer range. It was the most economical way of housing five or six cows, its total capacity, and so admirably suited to farms with few cattle. In particular, it was the plan used with field cowhouses in the Yorkshire Pennines, the rest of the building containing hay. Kilnside, Bishopsdale, is a late example of this plan used under one end of a barn, the threshing floor acting as the feeding passage (plate 8).

The fourth plan (fig. 34 c, d) had two rows of cows facing along the building, with a feeding passage between. Externally, it had three doors set fairly close together. It could be extended by repetition, but it then became effectively a series of separate cowhouses. The main advantages were economy of building, as it was smaller than the equivalent type 1, and ease of feeding, the passage being short. It was used fairly extensively in some areas in the eighteenth century. It went out of favour in the nineteenth, however, except in Cheshire and Lancashire, where it continued to be built up to about 1900. Type 5 (fig. 34e) is the previous one reversed, in that the cows face away from each other. Only rarely did it have feeding passages. Three areas where it is known to have been extensively used are the former county of Cumberland, the

Vale of Clwyd in Wales and north-east Scotland.

The final plan (fig. 34f) is found in two forms; the first and older is short and attached to another building, usually a barn. Examples can be seen in the former counties of Breconshire and Radnorshire and in north Cheshire and south Lancashire. The alternative was a long building, housing a considerable number of cattle; this seems to be a mid-nineteenth-century development.

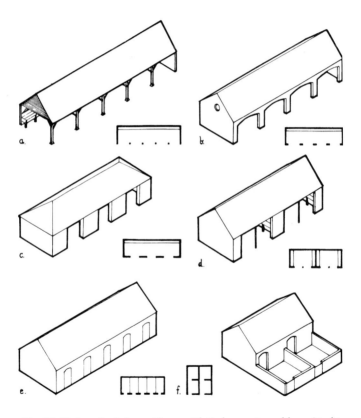

Fig. 35. Shelter sheds for cattle: a, with timber posts, gable omitted to show interior; b, with piers; c, with smaller openings, based on Cornish examples; d, with troughs across the building, based on Staffordshire examples; e, looseboxes, with timber partitions; f, hemmels.

Shelter sheds

The second type of building for cattle is the open-fronted shed, built to shelter loose cattle and related either to a yard or to a field (fig. 35). Normally the front wall was open, the roof being supported by piers or posts; occasional variations were for two or more sides to be open, or for the openings to be separated by a length of wall. Inside were a rack and trough as in the cowhouse, but there were no partitions or means of tying cattle, and the trough was set higher as the cattle fed standing. Only rarely was there a feeding passage, and the floor did not have a drainage channel. Usually the troughs and racks were along the back wall but occasionally the shed was divided up with cross walls to carry them (fig. 35d). This was designed to produce equal feeding and prevent the weak from being bullied. Most shelter sheds were single-storey; in Devon, however, they usually had an open-fronted loft above, forming a local type called a linhay (plate 26).

The shelter shed nearly always opened on to a yard. Where this was the main farmyard it was used for collecting manure from the cowhouse and stable, as well as for loose cattle. The sheds might extend to more than one side of the yard; alternatively the original sheds might appear very short for the size of the yard. This does not necessarily mean that there were only a few loose cattle to be housed, as the provision of shelter was sometimes rather inadequate. The ultimate in shelter for loose cattle was a completely covered yard, but this was only rarely provided before the late nineteenth century, except in Scotland. The advantages were that the manure was protected from the weather, less litter was required and the animals improved more quickly. These were not generally considered sufficient reasons, however, to justify the capital cost.

Looseboxes and hemmels

The loosebox was a room or pen in which the animal was free to move about. It might be large enough to take more than one animal but sometimes appears to have been too narrow for an animal to turn round properly. There was a trough for feed, set higher than in the cowhouse, and possibly also a rack. The loosebox was found in two forms. The usual form was one, or possibly up to three or four separate rooms in the farmstead. These were intended for calves, sick animals, a bull, or a cow about to calve. Where used for calves they might have access only from the cowhouse but otherwise usually had a single, external door. The alternative method of housing calves was in a series of small pens or stalls within the cowhouse. The second, much rarer form was a series of boxes divided by walls or movable wooden partitions (fig. 35e). They might each have an external door or form a series of pens within a large shed; they were built for fattening cattle. Cattle were considered to do better kept in boxes than tied in a cowhouse, but boxes were more expensive to erect and required more litter, and so were not often built.

The hemmel is a more expensive variation on the loosebox: the box is linked to an open yard of the same or slightly larger area, usually with an undoored opening between (fig. 35f). It was used for a bull or fattening cattle. Because it was expensive and needed even more litter than the loosebox it was only rarely built. In Northumberland by the early twentieth century this term was confusingly used for a type c shelter shed.

The dairy and slaughter house

The dairy is a room inevitably linked with the keeping of cattle. In it milk was converted to butter and cheese (the liquid milk trade was very restricted before the late nineteenth century). It normally formed part of the house and might adjoin a room furnished with racks in which the cheeses matured. Very occasionally it was a separate building and might then be quite ornate, as at Easton Park in Suffolk. Regulations introduced since 1885 have left few in their original form.

A slaughter house was rarely provided: it was generally built only when animals were killed regularly to supply a great house or because the farmer was also a butcher. It was high, to allow enough room to hoist the carcasses and for the wheel from which they hung. In more regularly used examples there would be a fasting pen adjoining; otherwise a normal loosebox might be used instead.

7. Buildings for other livestock

The pigsty

Pigs were found on most farms, being valued for their ability to fatten quickly and for eating what would otherwise be wasted. They were particularly connected with dairying, feeding on whey, that is the skimmed milk left over from butter and cheese making. Other food might be cooked in the kitchen or, less often, in a separate boiling house which adjoined the sty. The sties, for ease of feeding, were often near the house. Many farm and other country labourers kept their own pigs in a sty built in their garden: two such have been rebuilt at Blists Hill Open Air Museum at Telford, Shropshire.

The vast majority of pigsties in England and Wales are of one type, which had appeared by the early eighteenth century: a small box with a yard of the same or slightly larger size off it, the opening being without a door so that the animals could pass from one to the

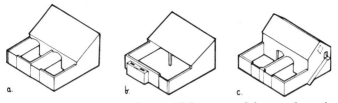

Fig. 36. Pigsties: a, normal type with lean-to roof; b, open-fronted, with chutes to trough; c, as type a, with loft for hens and hen ladder.

other at will (fig. 36a, c). The box was usually large enough for one or two fattening pigs or a sow and her litter. The feeding trough was in the yard, often with a chute for pouring in the swill from outside (fig. 36b). The building itself was low, reflecting the height of the pigs. Sometimes a hen loft was put above (fig. 36c), helping to keep both pigs and hens warm and making the sties more suitable for raising litters. The roof might be gabled or lean-to, the building detached or part of a larger range.

There are one or two variations. Occasionally the front wall of the box was omitted, as in a shelter shed for cattle (fig. 36b). Sometimes one or more boxes opened on to a large yard, which might be solely for pigs or shared with cattle. In some cases no separate buildings were provided for the pigs, who would shelter as best they could with the cattle.

There are two other types of sty, but they are generally not very common. The first is a series of rooms or boxes in a larger building with eaves at the normal height, usually with a feeding passage: access to the boxes may be only from that passage. In Devon, however, where this is the normal type, there is usually no feeding passage, and each room has its own external door. The second type is a fully roofed room with only a low external wall on one side, above which the side is open. This also usually has a feeding passage.

The fowl house

Hens were kept for their meat, eggs and feathers. Other large birds which might be kept were ducks, geese and turkeys. Sometimes no specific shelter was provided, the birds using the barn or cartshed, but this made finding the eggs a little difficult. Where accommodation was provided for them, it was basically one of two types, a loft or a room. The latter might be connected to a fowl yard.

The loft was probably used only for hens: it could be over pigsties (fig. 36c), the cartshed or livestock. It occupied the full roof space and if over a pigsty might, in addition, have low side

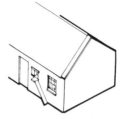

Fig. 37. Hen house, with ladder, based on a Northumbrian example, forming part of a longer range.

walls. When the sty was incorporated into a two-storey building, the hen loft might occupy the surplus space between the sty and the floor above. Such positions for the henhouse economised in building costs and when placed over livestock there was the advantage that the heat from the animals warmed the henhouse, encouraging laying in cold weather. Access to the loft was normally by a small door: keeping hens shut up all the time had appeared by the eighteenth century and is not just a modern practice! In many cases, however, the hens could get in and out through a small opening, about 12 inches (305 mm) high, approached by a ladder: this has now usually disappeared.

The room could be used for all four types of bird. Normally, there would not be more than one in the farmstead, forming part of a larger range of buildings as in fig. 37. Only occasionally would it be an independent structure or was a series of rooms built. An opening was left in the door or in the wall, about 3 feet (915 mm) up, with ladder access. Internally, there might be one or more rows of boxes on the walls for hens to lay in. A sloping rack might be built for hens and turkeys to roost on: geese and ducks needed a flat floor. Ventilation to rooms was usually from the door or window, but there might be ventilation holes; the last were particularly found in lofts.

The dovecote

Pigeons were originally kept to provide fresh meat but the importance of this declined with the increase in numbers of livestock by the eighteenth century; their dung was also useful. Later, during the nineteenth century, they may have been kept for their picturesque appearance. Perhaps because they fed themselves on any standing crops, the right to build a dovecote was strictly limited, initially to major landowners. Later, small freeholders could build one, and a tenant if he had his landlord's permission.

The best-known type of dovecote is a free-standing building,

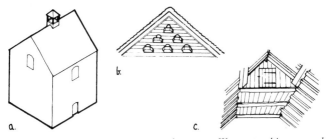

Fig. 38. The dovecote: a, seventeenth-century Worcestershire example with glover on roof; b, openings in gable to a small dovecote; c, small gable dovecote from inside.

built from the Middle Ages up to the late nineteenth century, the majority probably dating from before the eighteenth. The earliest seem generally to have been circular; others were square, multi-sided or rectangular. Some which appear rectangular were two square ones built together, as at Willington, Bedfordshire. Later examples were often designed to be ornamental; they might be first-floor rooms, with some other use below. Means of access for the birds varied, but a glover, an open-sided structure on the ridge, was probably the most usual (fig. 38a), followed by a window or dormer. The door is often not very obvious, being small and low so that the person entering filled the opening, preventing birds flying out whilst it was open. Internally there were rows of nests of stone, brick or wood, usually L-shaped. Access for cleaning the nests and catching young birds was by a ladder. This might be attached to arms projecting from a central post and could be revolved by the person on the ladder, so simplifying reaching all the nests. It was used in circular dovecotes, but experiment has shown that the corner nests in square dovecotes could also be reached from it. It was called a potence and may be a seventeenth-century invention; there is a good example at Dunster, Somerset.

Lofts were also used as dovecotes. The largest, like some built over barn porches, are really a variation of the first type. Normally, however, they were in the gable of a farm building or even the house (fig. 38b). These were usually small, restricted to the space above the upper purlins. Access externally was by one or two rows of holes, each with a small landing platform. Inside there was a door in the back. A third, rare alternative was for the nesting boxes to be on the external face of the building, usually the complete gable, with continuous landing platforms (plate 30). This type also covers those with only a few holes set under the eaves, as at the Ship Inn, Porlock, Somerset.

75

8. Recording farm buildings

In recording farm buildings, as in recording houses, it is advisable to work within a restricted area, beginning perhaps with a parish or a group of farms. This permits a complete picture to be built up: the area can always be enlarged later. It is also sensible to restrict the buildings examined by period: 1880 is a useful terminal date as it avoids the effects of the depression in arable farming which began then and the effects of the regulation of cowhouse design. It is also near the ultimate end of vernacular building techniques, although the vernacular content of the buildings had by then been declining for some time. An advantage of studying farm buildings is that the structure is usually exposed: walls are not often plastered, and only rarely does a ceiling hide the floor joists or roof trusses.

A clipboard, paper, pencil, tape-measure and camera are necessary, wellington boots and old clothes desirable. The first stage is to obtain the farmer's permission: owing to amalgamation of farms he may now live elsewhere, and the house may have been let out or sold, or the farm may belong to a company. Sometimes the buildings have been sold away from the land as well. Next a preliminary walk around the buildings will give an overall picture, very useful later in looking at them individually. It is possible to learn something about them by looking only at the outside, but this is a very restricting approach, unless your interest is only in building materials. To understand farm buildings it is necessary to go inside and to analyse, or try to analyse the plan. Whilst you may only record the original form of the building, it is very useful to note as well the alterations which fall within the period studied. These will help in understanding the development of the farmstead, and also to build up a picture, with other examples, of the development of the different types of building. The pigsties, goose pens and other lesser buildings should not be neglected, as they contribute to the complete picture.

Information from the farmer or an old labourer can be very useful in interpreting the buildings. They may be able to tell you how the cowhouse used to be arranged, what buildings have been demolished, what a building was like before the front wall was altered, and so on. Some farmers may have an aerial photograph of the buildings, useful in showing buildings which may have been demolished or altered, or yard boundaries which have been removed.

More can be learnt about the buildings and the farm from other sources. The county record office may have old maps, estate documents or sale catalogues. The last can be very useful, even if dating from after the end of the period studied, as they show how the buildings were used at the time of the sale.

A very useful guide to recording buildings is to be found in the *Illustrated Handbook of Vernacular Architecture* by R. W. Brunskill. The Historic Farm Buildings Group was formed in 1985 to study the subject; it publishes a journal and newsletter. The Vernacular Architecture Group publishes a *Bibliography of Vernacular Architecture*, listing current books and articles. General studies on farm buildings are: *History of Farm Buildings in England and Wales* by N. Harvey, *Traditional Farm Buildings of Britain* by R. W. Brunskill, *The English Model Farm* by S. Wade-Martins, and *English Farmsteads 1750-1914* by P. S. Banwell and C. Giles. *The Development of Farm Buildings in Western Lowland Staffordshire up to 1880* by J. E. C. Peters is a detailed regional survey and includes literary references which are applicable to other areas. There are sections on farm buildings in Brunskill's work noted above, in his *Traditional Buildings of Cumbria*, and in *Houses of the Welsh Countryside* by P. Smith. *Georgian Model Farms* (Oxford, 1983) by J. M. Robinson covers Great Britain. *Historic Farm Buildings* by S. Wade-Martins deals mainly with Norfolk. *Rural Settlement in Britain* by B. K. Roberts and *Village, Hamlet and Field* by C. Lewis, P. Mitchell-Fox and C. Dyer deal with settlement.

9. Places to visit

Listed here is a selection of farm buildings open to the public. Readers are advised to check opening times before travelling.

BUCKINGHAMSHIRE
Chiltern Open Air Museum, Newland Park, Gorelands Lane, Chalfont St Giles HP8 4AB. Telephone: 01494 871117. Website: www.coam.org.uk (Cruck barn, stable, granary on staddles.)

CAMBRIDGESHIRE
Wimpole Home Farm, Wimpole Hall, Arrington, Royston SG8 0BW (National Trust). Telephone: 01223 207257. Website: www.wimpole.org (1794 model farm designed by Sir John Soane.)

CORNWALL
Lanreath Farm and Folk Museum, Churchtown, Lanreath, Looe PL13 2NX. Telephone: 01503 220321. (Tithe barn.)

DEVON
Buckland Abbey, Yelverton PL20 6EY (National Trust). Telephone: 01822 853607. Website: www.nationaltrust.org.uk (Early-fifteenth-century century barn, late-eighteenth-century ox house, linhay.)
Newhall Equestrian Centre, Killerton, Broadclyst, Exeter EX5 3LW (National Trust). Telephone: 01392 462453. Website: www.nationaltrust.org.uk (Vernacular farmstead.)

PLACES TO VISIT

DURHAM

Beamish: The North of England Open Air Museum, Beamish DH9 0RG. Telephone: 0191 370 4000. Website: www.beamish.co.uk (*c.*1800 farmstead, horse engine house, some additional buildings.)

ESSEX

Grange Barn, Grange Hill, Coggeshall, Colchester CO6 1RE (National Trust). Telephone: 01376 562226. Website: www.nationaltrust.org.uk (Twelfth-century timber-framed barn.)

Priors Hall Barn, Widdington, Newport (English Heritage). (Aisled fifteenth-century barn.)

GLOUCESTERSHIRE

Ashleworth Tithe Barn, Ashleworth, near Gloucester (National Trust). Telephone: 01985 843600 (Regional Office) Website: www.nationaltrust.org.uk (*c.*1500.)

KENT

Wye College Agricultural Museum, Brook, Ashford TN25 5AH. Telephone: 02075 895111. Website: www.wye.ic.ac.uk (Timber-framed barn, hopkilns.)

LINCOLNSHIRE

Church Farm Museum, Church Road South, Skegness PE25 2HF. Telephone: 01754 766658. Website: www.lincolnshire.gov.uk (Nineteenth-century farm buildings.)

OXFORDSHIRE

Cogges Manor Farm Museum, Church Lane, Cogges, Witney OX8 6LA. Telephone: 01993 772602. (Large farmstead with buildings dating from the sixteenth century to the late nineteenth century.)

Great Coxwell Tithe Barn, Great Coxwell, near Faringdon SN7 7LZ (National Trust). Telephone: 01793 762209. Website: www.nationaltrust.org.uk (Thirteenth-century aisled stone barn.)

Swalcliffe Barn, Shipston Road, Swalcliffe, near Banbury. Telephone: 01295 788278. (Large fourteenth-century stone barn.)

SHROPSHIRE

Acton Scott Working Farm Museum, Wenlock Lodge, Acton Scott, Church Stretton SY6 6QN (Shropshire Museum Service). Telephone: 01694 781306. Website: www.shropshire-cc.gov.uk/museums.nsf (Model farmstead of 1769.)

SOMERSET

Dunster. (A good medieval dovecote with a potence, by the church.)

Somerset Rural Life Museum, Abbey Farm, Chilkwell Street, Glastonbury BA6 8DB. Telephone: 01458 831197. Website: www.somerset.gov.uk (Fourteenth-century stone tithe barn.)

West Pennard Court Barn, West Pennard, near Glastonbury, Somerset (National Trust). Telephone: 01985 843600 (Regional Office). Website: www.nationaltrust.org.uk Please note that visits are by appointment only, to arrange call 01458 850212. (Fifteenth-century stone barn)

STAFFORDSHIRE

Staffordshire County Museum, Shugborough, Milford, Stafford ST17 0XB. Telephone: 01889 881388. Website: www.staffordshire.gov.uk/shugborough (Nineteenth-century home farm.)

SUFFOLK

Easton Farm Park, Easton, near Wickham Market IP13 0EQ. Telephone: 01728 746475. Website: www.eastonfarmpark.co.uk (Model farmstead, *c.*1870, with an ornate dairy.)

Museum of East Anglian Life, Abbot's Hall, Stowmarket IP14 1DL. Telephone: 01449 612229. Website: www.suffolkcc.gov.uk/tourism (Medieval barn.)

SUSSEX, WEST

Weald and Downland Open Air Museum, Singleton, Chichester PO18 0EU. Telephone: 01243 811363. Website: www.wealddown.co.uk (Barns, shelter sheds, cartshed, granary and stable dating from the seventeenth to nineteenth centuries.)

WARWICKSHIRE

Mary Arden's House, Station Road, Wilmcote, near Stratford-upon-Avon CV37 9UN. Telephone: 01789 293455. Website: www.shakespeare.org.uk (Barn and dovecote.)

WILTSHIRE

Bradford-on-Avon Tithe Barn, Barton Farm Country Park, Pound Lane, Bradford-on-Avon BA15 1LF (English Heritage). Website: www.english-heritage.org.uk (Large medieval barn.)

WORCESTERSHIRE

Avoncroft Museum of Historic Buildings, Stoke Heath, Bromsgrove B60 4JR. Telephone: 01527 831363. Website: www.avoncroft.org.uk (Barn, stable, granary and cartshed, *c.*1800, dovecote.)

Bredon Barn, Bredon, near Tewkesbury (National Trust). Telephone: 01743 708100 (Regional Office). Website: www.nationaltrust.org.uk (Aisled stone barn *c.*1340.)

Middle Littleton Tithe Barn, near Evesham (National Trust). Telephone: 01743 708100 (Regional Office). Website: www.nationaltrust.org.uk (Thirteenth-century stone barn.)

YORKSHIRE, NORTH

Newham Grange Leisure Farm, Wykeham Way, Coulby Newham, Middlesbrough TS8 0TG. Telephone: 01642 300261. Website: www.middlesbrough.gov.uk

YORKSHIRE, WEST

East Riddlesden Hall, Bradford Road, Keighley BD20 5EL (National Trust). Telephone: 01535 607075. Website: www.nationaltrust.org.uk (Two seventeenth-century barns.)

Shibden Hall and Folk Museum of West Yorkshire, Shibden Hall, Listers Road, Halifax HX6 6XG. Telephone 01422 352246. Website: www.calderdale.gov.uk (Seventeenth-century barn.)

WALES

Museum of Welsh Life, St Fagans, Cardiff CF5 6XB. Telephone: 029 2057 3500. Website: www.nmgw.ac.uk (A number of longhouses and other Welsh farm buildings.)

Index